Horizons *of* Cosmology

Templeton Science and Religion Series

In our fast-paced and high-tech era, when visual information seems so dominant, the need for short and compelling books has increased. This conciseness and convenience is the goal of the Templeton Science and Religion Series. We have commissioned scientists in a range of fields to distill their experience and knowledge into a brief tour of their specialties. They are writing for a general audience, readers with interests in the sciences or the humanities, which includes religion and theology. The relationship between science and religion has been likened to four types of doorways. The first two enter a realm of "conflict" or "separation" between these two views of life and the world. The next two doorways, however, open to a world of "interaction" or "harmony" between science and religion. We have asked our authors to enter these latter doorways to judge the possibilities. They begin with their sciences and, in aiming to address religion, return with a wide variety of critical viewpoints. We hope these short books open intellectual doors of every kind to readers of all backgrounds.

Horizons *of* Cosmology

EXPLORING WORLDS SEEN AND UNSEEN

Joseph Silk

TEMPLETON PRESS

Templeton Press
300 Conshohocken State Road, Suite 550
West Conshohocken, PA 19428
www.templetonpress.org

Designed and typeset by Gopa and Ted2, Inc.

Library of Congress Cataloging-in-Publication Data

Silk, Joseph, 1942–
Horizons of cosmology : exploring worlds seen and unseen / Joseph Silk.
p. cm.
Includes bibliographical references and index.
ISBN-13: 978-1-59947-341-3 (alk. paper)
ISBN-10: 1-59947-341-0 (alk. paper)
1. Dark matter (Astronomy) 2. Expanding universe.
3. Cosmology. I. Title.
QB791.3.S55 2009
523.1—dc22

2009010014

Printed in the United States of America

09 10 11 12 13 14 10 9 8 7 6 5 4 3 2 1

Color image credits: Series A–E: STSci; G1: 2DF Galaxy Redshift Survey, Anglo-Australian Observatory, 1999; G2: NASA/WMAP Science Team; G3, G4: V. Springel et al., Max-Planck Institut fur Astrophysik, Garching, 2005.

To the next generation:

Timothy, Jonathan, Edouard, Cyril, and Jonah

Contents

Preface: A Wondrous Place ix

Chapter 1: Cosmology Begins 3

Chapter 2: Case for the Big Bang 16

Chapter 3: Inflation Explained 38

Chapter 4: How Stars Form 52

Chapter 5: The Darkest Matters 63

Chapter 6: Cosmic Archaeology 74

Chapter 7: Detecting Dark Matter 102

Chapter 8: Finding Dark Energy 118

Chapter 9: Eminent Missteps in Cosmology 132

Chapter 10: The Universe in Seven Numbers 149

Chapter 11: Our Place in the Universe 160

Chapter 12: Cosmology's Future 178

Glossary 187

Bibliography 193

Index 197

PREFACE
A Wondrous Place

THE UNIVERSE is a wondrous place. Its marvels surpass human imagination. Stars form and die in incredibly colorful displays. The chemical elements are cooked in the hot interiors of stars and dispersed in immense explosions. Vast numbers of stars pirouette in giant galaxies. Immensely massive black holes seed huge outflows that illuminate the remote depths of the universe. All of this and more constitute the catalogue of objects that populate the universe.

Astronomical observations demonstrate that the universe began as a dense fireball. As it expanded and cooled down, matter and eventually massive gas clouds condensed out. Galaxies and stars formed. Ninety percent of the universe is dark, and not made of ordinary matter. The universe will continue to expand, essentially forever. Such was our past, and such will be the future, as revealed by the wonders of modern astronomy.

It is remarkable how much we have learned about our universe in the past century. When Albert Einstein said that "it should be possible to explain the laws of physics to a barmaid," he was only half joking. We do have a fairly simple explanation of our universe. But behind that simplicity lurks a great deal of mystery and complexity. Cosmology is filled with puzzles and speculations. I have tried to tell this story simply enough for general readers, while also trying to retain enough technical detail for advanced students and professionals.

With that in mind, this book roughly falls into three parts. It begins by narrating the basic story of our discoveries in cosmology over the past century—an expanding universe populated with galaxies and stars. The middle chapters look at a number of contemporary puzzles. These puzzles relate especially to galaxy formation and to the jarring new reality of dark matter and dark energy. All of these topics are of particular interest to cosmologists today. These middle chapters lean toward a more technical discussion. The reader will have noticed by this time that cosmology is a constant back and forth between observation, theory, and trying to test those theoretical predictions.

The last third of the book looks at the human side of cosmology and moves to the more speculative and philosophical frontiers with which cosmologists are also wrestling. Einstein once declared, "The man of science is a poor philosopher." Still, when talking about the origins, nature, and future of the cosmos—or the concepts of ultiverses and time travel—cosmologists unavoidably cross the line into speculation.

I have been privileged to be a witness to many of the discoveries in cosmology since the 1960s, and have tried my own hand at several specific problems. Cosmology began with a few lone figures who had arrived at some of the greatest insights about the structure and dynamics of our universe. It now is a collective project. There are too many notable people in the field to name them all in these pages. In this new century, we all face the challenges of finding the budgets necessary to build larger instruments to test our best theories. However, as I say at the end of this book, we are not necessarily in a great hurry. If all goes well, we have many centuries ahead in which to find out the true nature of the cosmos around us.

Horizons *of* Cosmology

CHAPTER 1
Cosmology Begins

MODERN COSMOLOGY began in the 1920s when two men with different backgrounds, and about as separate geopolitically as could be imagined, entered the scientific scene. Both shared a love of physics and a fascination with Einstein's recently launched theory of gravity, which said that on the largest scales of the universe, space-time was like a curved fabric that was being pulled inward. The great problem for Einstein's mathematics, and the debate that followed, was why the universe looked stable when it should have been collapsing from the force of gravity.

This is where the two founders of modern cosmology, Russian physicist Alexander Friedmann (d. 1927) and Belgian physicist Georges Lemaître (d. 1966), started to make their revolutionary proposals about the universe.

Friedmann was a meteorologist, a military pilot in World War I, and a professor of mathematics and physics in the Soviet Union. He died prematurely of typhoid within a month of setting the world altitude record for a balloon flight in 1925 at seventy-four hundred meters. But three years earlier he made a major discovery regarding mathematical problems in the field equations of Einstein's theory of gravity. Friedmann had studied Einstein's theory of general relativity in detail, and was one of the very few who understood it. Friedmann discovered what we call today the expanding universe solution to Einstein's equations, which explains why the universe is not collapsing, despite the implications of Einstein's general relativity theory of gravity.

The Russian corresponded with Einstein, who promptly criticized Friedmann's result as erroneous. But Einstein retracted his attack the following year, and Friedmann became famous, at least in Russia. Einstein eventually realized he had made an elementary mathematical mistake and missed the new solution.

In the 1920s, when communications between scientists in Russia and the west were hardly optimal, Lemaître also began to look at the cosmological implications of Einstein's theory of gravity. He, too, arrived at a solution, coming a bit after Friedmann but not knowing the Russian's work, some years later. In Belgium, Lemaître was a Roman Catholic priest. He eventually became president of the Pontifical Academy of Sciences and the principal science advisor to Pope Pius XII. Lemaître began this rise in his field with a graduate studies fellowship at the Massachusetts Institute of Technology. His American visit was also marked by a trip to the California Institute of Technology, a visit crucial to Lemaître's later theories about an expanding universe.

Pasadena was home to the small organization (funded by the Carnegie Institution) that operated the world's largest telescope, which had a one-hundred-inch-diameter mirror and sat atop Mount Wilson, east of Los Angeles. On his California trip, Lemaître met cosmologist J. P. Robertson and also Edwin Hubble (d. 1953), the chief observational astronomer at Mount Wilson. Hubble was not interested in theory the way that Robertson was. But with his Mount Wilson telescope, he was experimenting with the measurement of redshift, which refers to the way the light of distant galaxies moves toward the red end of the spectrum when those galaxies are moving away from the observer.

Lemaître got a firsthand look at the difficulties of measuring redshifts. With this new kind of information in tow, he returned to Belgium and began his attack on the problems in Einstein's theory of gravity. In 1925, quite independently of Friedmann, Lemaître also discovered the expanding universe solution to Einstein's equations. Lemaître formulated a law that he believed would relate redshift and the distance of galaxies to the expanding universe. With this idea, Lemaître had conceived of the first modern cosmological test: a theory that could be measured by observational data.

In the meantime, back in Pasadena, Hubble was developing his own interpretation based on his observations. In 1929, he came up with what is now called Hubble's law, which asserted that the recession rate of the galaxies is linearly proportional to distance. In short, galaxies move faster as they are farther away from the observer, which also was a prediction of, and evidence for, the expanding universe theory. The story of Hubble is itself an early account of American astronomy.

Hubble had studied mathematics and astronomy at the University of Chicago, but he also excelled in boxing. He turned down an offer to train for the world heavyweight championship in favor of a Rhodes scholarship at Oxford where he obtained a law degree. On returning to the United States, he practiced law for a while but returned to his first love, astronomy. He obtained his doctorate at the Yerkes Observatory, operated in Wisconsin by the University of Chicago. Then, in 1919, he joined the staff at the Mount Wilson Observatory. His timing was fortuitous. Two years earlier, the one-hundred-inch telescope had begun operation.

Hubble exploited the new telescope to explore the frontiers of the universe. He resolved individual variable stars in the spiral nebulae and demonstrated that they were Milky Way–like galaxies, which astronomers of that period referred to as island universes. The eighteenth-century philosopher Immanuel Kant had speculated that island universes are distant galaxies, but it was only wishful thinking without the data to support this viewpoint. Indeed, no physicist, especially Einstein, had anticipated that the Milky Way was only a minor and inconsequential constituent of the universe, surrounded by uncountable numbers of neighbor galaxies.

Edwin Hubble developed one of the most unusual partnerships in modern science. His accomplice was a former mule driver and Mount Wilson janitor, Milton Humason, who began work as a night assistant for the telescopes and eventually was hired as Hubble's chief assistant. Humason, who had left school at fourteen, became one of the world's most skilled astronomers. Humason's expertise was in using the telescope to take photographs of galaxies and develop them at sensitivities that no one has previously attained. It was this skill that played an important role in his and Hubble's joint quest to identify and measure the most distant galaxies.

As is common in astronomy, one discovery builds on another. Hubble's new distances attained significance only after being combined with data obtained by Vesto Slipher, who in 1915 had first reported that most nearby galaxies were receding from us. And it took some years before all the best theories and data—from Friedmann, Lemaître, Hubble, and Slipher—were being looked at together.

At the now-historic Lowell Observatory in Flagstaff, Arizona, Slipher had used a twenty-four-inch refractor telescope. It was painstaking work. Each exposure took up to forty hours. He used the technique of spectroscopy, which relies on the Doppler effect, well known by how the sound waves of a train whistle change when it moves away from the observer. Slipher used this same approach to measure the radial velocity of galaxies, that is, the components of motion away from us. Indeed, he found that entire galaxies moved predominantly away from the earth. The light that reached his telescope, called spectral lines, shifted to longer wavelengths, again showing the redshift effect. However, without knowing accurate distances for the galaxies, Slipher could not possibly deduce much more than a large systematic motion of the apparently nearby objects.

During the 1920s, in fact, all the evidence for an expansion of the universe remained conjectural, even though it was a topic of lively debate, orchestrated most notably by Knut Lundmark in Sweden, as well as by others. As always, the weak link remained the uncertainty in nebular distances. Hubble broke that stalemate in 1929 when he announced his correlation between the distance to a galaxy and its redshift, now called Hubble's law. This was promptly interpreted as a measure of the velocity of the galaxy away from the observer. Most but not all galaxies were receding, Hubble announced.

Lemaître jumped on the new data. Amid the 1920s speculation, it was proposed that the universe expanded from a singular point of origin in time. The concept of an initial infinitely dense singularity was an idea Lemaître could not accept. The great cosmologist Yaakov Zel'dovich, speaking later on this choosing of a model of origins, famously joked that "the point of view of a sinner is that the church promises him hell in the future, but cosmology proves that the glowing hell was in the past"—at the first fiery explosion.

Motivated by a more metaphysical logic, Lemaître instead proposed that the universe began its expansion from a finite size. He dubbed it the primeval atom. He supported the idea of a tiny initial universe by adopting Einstein's mathematical trick, which was to introduce a cosmological constant—a hypothetical force—that holds back the universe from collapsing under its own gravity. While Einstein had lost interest in this antigravity constant once it was clear that the universe expanded, Lemaître revived the idea. He used the constant to explain how his finite initial state, the primeval atom, had managed to exist at the beginning. One advantage of Lemaître's model was that it could account for the age of the universe. In the absence of a cosmological constant, Hubble's method dated the universe as being even younger than the earth.

Reluctant at first, Hubble finally conceded that his 1929 discovery meant that the universe might indeed be expanding via the recession of the galaxies. The evidence began to seem overwhelming despite the fact that Hubble's first approach gave an unacceptably young age for the universe. Hubble explained his change of mind in his 1936 book, *The Realm of the Nebulae*: "Redshifts resemble velocity shifts, and no other satisfactory explanation is available at the present time: redshifts are due either to actual motion of recession or to some hitherto unrecognized principle of physics."

By 1942, the recession velocities measured by Hubble and his colleagues had reached speeds of up to twenty-five thousand miles per second, or one-seventh the velocity of light. These rates were enormous. But even earlier, Hubble could not accept the image of a rapidly expanding universe. His skepticism obliged him to choose the other alternative stated in his book: that the light in the universe operated under some new principle of nature, such as might be contained in some as yet undiscovered new physics.

At this point, Hubble proposed the idea of tired light to explain the redshift. This hypothesis represented a fundamentally new physics, though Hubble was not a physicist by training. To give him credit, Hubble did realize that the discovery of tired light would be as great a revolution in physics as was the expanding universe, and he was fully aware of the lack of any contemporary evidence for this phenomenon. The argument that photons lose energy traveling across space, hence shifting to the red, has now been definitively discarded on experimental grounds.

More to the point was a demonstration of the reality of the expansion. To convince himself and others of the reality of the expansion, Hubble devised a new test. In an expanding space, the phenomenon of time dilation occurs. If distant space is expanding, the rate of arrival of photons from those distant sources must decrease. It is the perfect test, though the results came in long after Hubble's heyday. The test was implemented in 1990 by measuring the light curves of the brightest known objects in distant galaxies, which are Type 1a supernovas (star explosions). The current theory is that all supernovas, as perfect bombs, emit the same rate of light. Once this light is corrected for time dilation, supernovas that are nearby *and* distant all show identical light curves. This proves time dilation and hence the expansion of the universe.

Hubble's erroneous age of the universe was due to the fact that his constant was far too large. Why was Hubble so far off the modern value for the expansion rate? Hubble's original diagram seems largely wishful thinking to the modern observer. The spread in his data is enormous. After all, the inferred value for Hubble's constant (some six hundred kilometers per second per megaparsec)[1] was too large by a factor of ten. However, his later data vindicates, if not the modern value, certainly the underlying trend of the expansion law with distance that was caught by Hubble's inspired intuition.

A vigorous debate continued in the post-Hubble decades. Rival schools argued for a low or a high value of Hubble's constant, considered to be either fifty or one-hundred kilometers per second per megaparsec. Neither view was grossly wrong, although it required the aptly named Hubble Space Telescope to settle the controversy. Today, the accepted value of the Hubble constant is seventy-two kilometers per second per megaparsec, with an uncertainty of about 10 percent. This gives us a universe that is 14 billion years old, comfortably older than the ages of the oldest stars.

With all of these advances, there was no reason to go back to Lemaître's cosmology of the primeval atom. First of all, the primeval atom would be unstable and collapse, which would be a disaster for the early universe. Another strike against the primeval atom theory came later in the twentieth century, when it was discovered that the entire universe had an evenly spread heat, what we today call the cosmic microwave background, which proves to be the ovenlike warmth left over from a hot big bang beginning of the universe. This was the death knell for any theory that lacked a dense and hot beginning. But for the big bang theory to win the day, it too had to gradually build up its evidence, explaining the data, as we see in the next chapter.

1. The parsec, a distance of about 19.2 trillion miles, or 3.26 light years, is a basic measure for galaxy distances. A kiloparsec is a thousand parsecs, and is typically used to measure between parts of galaxies or in groups of galaxies. A megaparsec is a million parsecs. It typically is used to measure between different galaxies or different clusters of galaxies.)

Galaxies Rule

As illustrated by these debates in early cosmology, the galaxies were the crux of the matter. To the jaundiced eye of a career astronomer, they are still the most beautiful objects in the universe. They are useful for science as well. Galaxies are beacons that illuminate the universe. We use galaxies as markers to chart space. We can use them to probe the evolution of the universe. In more recent years, they are helping us measure the effects of the greatest new puzzle in physics: the existence of dark matter and dark energy, which we look at later in greater depth.

There are more than 10 billion galaxies in the observable universe. In turn, the average galaxy contains more than 100 billion stars. Although galaxies come in many shapes and sizes, they are all essentially diffuse gas and dust clouds with volumes equaling millions of solar masses (a measure based on the mass of our sun). These gas clouds are the birthplaces of the stars. We also see stellar death. The full life cycle of the stars can be studied. As in any large population, in which there are people of all ages, so there are stars at all stages of evolution. Just as in a human population, studying birth and death yields insights into the nature of the stars.

At least a few percent of the stars are surrounded by planetary systems. This adds up to a lot of planets, most of which are too far away to be of much interest. Still, the discovery of planets may be one of the most important for the future. Many of these planets are thought to be earthlike. Whether they harbor life in any form is completely unknown and a matter of continuous speculation. Maybe with enough earthlike planets, life is inevitable. Or maybe it is not. Only future observation may tell.

In these pages, we often refer to three shapes of galaxies: a flat disk, a spiral disk with a central bulge, or an ellipse (oval) that is pure bulge. Nearly all galaxies have a bulge, like a soccer ball, at their center. This is called a galactic spheroid, a term also used when an entire galaxy bulges. A final element in the shape of galaxies is an elongated bulge called a bar. Our galaxy, the Milky Way, for example, is a spiral disk galaxy with both a galactic spheroid and a bar at its center.

Once a galaxy takes on a final shape, it reveals certain characteristics. Spiral galaxies contain many young stars, for example. These often brilliant and hot, and hence blue, objects dominate the light from a spiral galaxy and make the spiral arms visible to us. Elliptical galaxies contain only older stars. Their light is red, the characteristic color of stellar aging.

Some galaxies are like train wrecks. They are unclassifiable, and are often found to be merging galaxies. Mergers provide a glimpse into a more violent past, as they were once far more frequent in the universe. Elliptical galaxies form by mergers.

Spiral galaxies contain large amounts of diffuse gas. This interstellar matter is the raw material for forming stars. We observe prolific star formation in spirals. We also see the aftermath of formation, which is stellar aging. Spirals form by accumulating gas over billions of years.

Galaxies seem to form inside out, starting with an original spheroid at their centers. The oldest stars are in these central regions. These older stars are all red, but the oldest ones are redder because they contain more metals. Over their longer histories, the reddest stars have collected heavy elements, especially iron, that have been recycled by previous generations of stars. Stars with high metal content are more abundant in the inner parts of galaxies, so we know that these regions formed first. The colors give us a precise means of dating star and galaxy formation.

Hubble's Types

Edwin Hubble specialized in observing galaxies, and thanks to his early work, we have developed a system to sort and classify galaxies (called a phenomenological approach to astronomy, distinct from the theoretical approach). Hubble found that galaxy images could be arranged in a sequence that progresses from galaxies dominated by spiral arms to those in which spiral arms are a subdominant feature. These were all flattened systems, or disks. At the end of the morphological sequence were the ellipticals—featureless ellipsoidal systems whose extent of flattening varied from being completely round to being flattened by about 50 percent. The first question one might pose is whether the morphological sequence is also a sequence of evolution. And if so, what is the direction of the evolution?

The first clues came from recognizing the different types of stars and their inferred ages. Stellar ages are obtained once the luminosity and color of a star are measured. Most stars are burning hydrogen, the principal fuel source. Luminous blue stars burn their nuclear fuel rapidly, and are hot and short-lived. Dim red stars are cool and are long-lived. There are also short-lived luminous red stars: these are the red giants and supergiants, the fate of stars once the core hydrogen fuel is exhausted. And there are dim blue stars: these are the final stages of evolution of the red giants into white dwarfs. A galaxy is a mix of stars of all ages. We cannot resolve individual stars for most galaxies, so we study the collective light in a resolution element that might contain millions or billions of stars, depending on the distance to the galaxy.

The light from a galaxy consists of the collective emission from its stars. The light provides information about the average age of the stellar population and its chemical abundances. Disk galaxies are found to be blue, ellipticals to be red, which means that the stars that dominate the integrated light from disks are young, tens of millions of years old. Conversely, the stars in ellipticals are old, typically like the sun or even older, and which means at least several billion years old. In fact, disk galaxies also contain older stars. They have undergone a prolonged star formation history, including stars that are forming at present. These young stars are the hottest, bluest, and most luminous, and dominate the light.

Cosmic Evolution

The galaxies tell us how the universe evolved, past to present. Look deep into the universe, and one looks back in time. The most remote galaxies are necessarily youthful. Galaxies age as they form enough stars to exhaust the initial gas supply. By taking large samples of galaxies, we can study their evolution as a consequence of aging. In aging galaxies, chemical elements are more abundant, and the galaxy morphology changes. Young galaxies tend to be irregular in shape and are undergoing high rates of star formation. Today's universe is more ordered.

The history of the universe is also told by black holes, points of immense gravity (called black because even light cannot escape the gravitation). Supermassive black holes are found in the centers of nearly all galaxies. Now they are mostly inert, but long ago they played a far more active role as sources of extremely energetic phenomena. Quasars, described later, are the most luminous objects in the universe and are one kind of active nucleus of galaxies. In the early universe, these active galactic nuclei were more frequent—powered, we believe, by more active supermassive black holes. The immense gravity of the black holes heated up any gaseous disks that formed nearby, producing immense explosions. (We return to black holes in a later chapter.)

The violent phenomena that emanated from the central active nucleus of a galaxy had a dramatic impact on young galaxies. Today, using spectroscopy, we can see the behavior of heated and energized gas, which can be seen flowing at velocities up to a tenth of the speed of light. This violent movement of gas can interrupt the gas supply that allows for star formation, for example. Hence, these energetic nuclei, by their sheer violence, can terminate a galaxy's gas-rich phase of prolific star formation.

Nevertheless, star formation goes on. Other stars are continually shedding gas and debris as they evolve or explode. Interstellar clouds gather up the debris from previous generations of stars. Larger clouds also draw in the gas of smaller clouds, finally forming a cold disk that, in turn, breaks up into clumps of gas—the seeds for new star formation. Today's galaxies are a mixture of star-forming disks, characterized by beautiful spiral structures that demarcate the star-forming clouds, and old, red spheroidal galaxies, where star formation has mostly ceased. Rejuvenation, or birth of new stars, is by no means inevitable. It depends on the environment. In the great clusters of galaxies, the pressure of the hot intergalactic medium strips out interstellar gas, and the majority of cluster galaxies are spheroidal galaxies.

In addition to revealing how the universe evolved, galaxies also have their own life spans. They are born, age, and die. We are particularly interested in their aging process. We monitor aging by studying the stars' chemical abundances, which build up with time. Stellar explosions are the prime source of the heavy elements. More and more debris from exploding stars is incorporated into new generations of stars. Thus we look in the outer regions of a galaxy where there is less time for recycling. Over the greater distances involved, stars should be more metal-poor and older than their counterparts in the inner regions of the galaxy, which is exactly what we observe.

This recycling of material is remarkably efficient. We can even say that the universe is "ecologically correct," in that the recycling operates in the interstellar gas (gas between stars), which contains the debris of many generations of stars. Some of this gas is stripped from the galaxies and resides in the intergalactic medium (the debris between galaxies). Studying intergalactic clouds provides a way of probing the material from which the galaxies formed. The galaxies shed gas as well by the cumulative effect of many dying, exploding stars. Over many billions of years, the intergalactic gas is progressively enriched, always lagging behind the stars in terms of heavy element content. This gas is the reservoir for galaxy and star formation recycled into future generations of intergalactic clouds.

The process has allowed astronomers to probe the evolution of the universe as well. Looking back in time, it is predicted that the recycling was progressively less, and therefore galaxies are progressively less enriched with chemical elements. They had less time to recycle and accumulate the debris of other dying stars, exactly what we have observed with the aid of the world's largest telescopes.

The death of stars is central to the cosmic recycling process. All of the chemical elements in the universe are the ashes of long-dead stars. Thanks to our telescopes and our mathematical physics, we have studied this process of stellar death in great detail. When we see red giant stars, for example, we know that these are low-mass stars that have run out of hydrogen fuel. These stars burn very brightly: they are very large and very red. Our own sun will go through this phase in about 5 billion years—and incinerate the earth in the process. We on earth must look forward to this inevitability, but there will doubtless be enough time to find an escape route.

For slightly more massive stars than our sun, the central temperature reaches the point at which helium will burn. The star becomes a red supergiant. These stars are so bright that we can detect them in distant galaxies. The outer layers of the star are shed by the pressure from the intense heat generated as helium burns. A shell of glowing gas surrounds the dying star. These are some of the most beautiful objects in the universe. In the nineteenth century, these fuzzy images were confused with planets, so today we still call this evolutionary stage of single stars a planetary nebula. They are also the source of most of the carbon in the universe, essential to biological life on earth. The images taken with the exquisite resolution of the Hubble Space Telescope reveal a network full of threads and sheets of glowing gas. A compact, inert mass of carbon and oxygen is left behind—a white dwarf that cools into invisibility over billions of years.

The various abundances of the elements in our universe are like fossils to the archaeologist, helping us trace our cosmic evolution from the past. This entire process has not only produced planets, but also carbon, the key element for life. We are literally descendants of the stars.

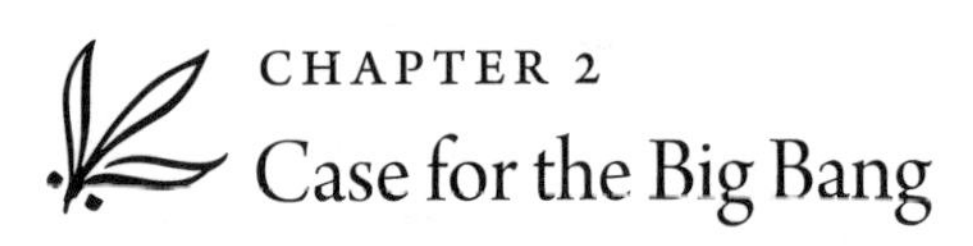

CHAPTER 2
Case for the Big Bang

THE IDEA of an expanding universe was a shock to early astronomers, but now the jury is in: the universe is indeed expanding. This is the inevitable consequence of Hubble's observations. Curiously, Hubble himself never accepted such a paradigm shift although it stemmed directly from his work. He rather chose to accept galaxy redshifts as an observable phenomenon without any commitment as to their origin in terms of the properties of space. Perhaps he was confused by the early cosmologists, pre-Friedmann, who favored the cosmology model of Dutch astronomer Wilhem de Sitter. In 1917, de Sitter proposed a static universe in which a hypothetical field produced the observed redshift. A de Sitter universe is actually devoid of matter and not expanding.

The systematic recession of the galaxies is now explained as being due to the expansion of space. Einstein's theory of gravitation certainly predicts this phenomenon. But why is space expanding? This question takes us back to the initial conditions of an infinitesimal patch of matter from which the universe began. That matter must have been in a volatile state, that is, out of equilibrium. This could have been a state of contraction or of expansion. Either way, the density of this primordial patch must have been 10^{90} grams per cubic centimeter. This is known as the Planck density, after German physicist Max Planck. This density is so high that it takes place only at the interface of quantum theory (in tiny atoms) and general relativity (large-scale gravity). In other words, at the initial conditions, the smallest and largest forces in the universe known today were squeezed together, united and indistinguishable.

The quantum processes were operating in the patch. By quantum jumps, macroscopic clumps of matter could disappear or reappear like the Cheshire cat in *Alice in Wonderland*. Black holes, which are so dense with gravity that they attract all the matter around them, could have formed and decayed spontaneously. In this early state, the universe must have been at the most extreme density that can be conceived under known physics. It represents our best guess at the conditions that prevailed near the beginning of time.

After that, the direction of the universe has been quite predictable. It has expanded according to our basic measuring tool, the Hubble diagram, which plots distance compared to the velocity of galaxies as they move away from the central starting point of the universe. We deduce that this expansion began 13.7 billion years ago. The latest data, using supernovae to chart the expansion, have added something surprisingly new to the traditional Hubble diagram: the remotest galaxies are accelerating in recession, speeding up the expansion of the universe, a topic we discuss later.

The ancient age of the universe has also been a surprise to modern science, at least for a century or so. Today, scientists subscribe to the view of a very old universe of about 14 billion years. It is a difficult idea for a substantial minority of the population, especially in North America. Many people prefer a traditional interpretation of the universe drawn from a literal reading of the Bible. In one famous calculation from the King James Bible by seventeenth-century Anglican bishop James Ussher, the universe was created in 4004 BC on Sunday, October 23, at about 7:30 a.m. Today, decades of Gallup polls show that up to 50 percent of Americans think that human life arose fairly recently, according to a literal reading of Genesis, and for many, this would also include the belief in a very young universe.

Fortunately, from the time of Pope Pius XII in the 1940s, guided by the advice of astronomers such as Abbé Lemaître, the Catholic

Church and other religious circles have taken a more enlightened approach to modern cosmology, which tries to find a proper balance between theology and science. This view holds that while science is paramount, it presents no challenge to a creed that rests on beliefs that arise from faith. Indeed the converse also applies: the beauty of science and the revelations produced by scientific discovery constitute part of the modern theologian's perspective and toolbox.

Today, for example, the discoveries of modern physics, astronomy, and cosmology reveal intricate details in the physical structure of the universe that seem highly improbable. The proton mass is remarkably close to the neutron mass. Were it very different, stars would not have formed. Further, the force that is accelerating the universe is far weaker than physics leads us to expect. Were this force much stronger, galaxies would never have formed. And in a universe devoid of stars and galaxies, there would not be any observers to marvel at the mysteries of the cosmos. It is not hard to see how theologians might find such discoveries fascinating.

These apparent coincidences in the universe have prompted some to argue that the arrival of human beings on earth is perhaps not a cosmic accident after all. Indeed, those who employ this reasoning have elevated this human-centered argument into a fundamental principle that governs the universe, which has now been called the anthropic principle, for *anthropos,* or man. This principle has long held sway in traditional religion. But sadly, in the view of some, the wheel has turned full circle and now physicists too are appealing to the anthropic principle to account for the initial conditions of the big bang. Obviously, the anthropic approach is an unabashedly self-based egocentric worldview—a topic we explore at the end of this book.

Following the Evidence

Our concern now is the evidence for the big bang theory of the universe, for we do not want to take it just on hearsay. Four major predictions of the big bang theory have been verified by modern scientific experiments: the recession of galaxies, the abundance of light elements in the universe, the existence of a cosmic background radiation (blackbody) that is uniform, and finally predicted rates of fluctuations in that same radiation. These four lines of evidence ought to be enough to quench even the most biased critics of what at first sight is a highly implausible theory.

As we recall, Friedmann and Lemaître had predicted an expanding universe, which Hubble then measured. What Hubble pioneered and refined was the correlation between recession velocities and the distances to remote galaxies. We judge the velocities by the spectra of light in galaxies. The redshift of spectral lines shows the recession velocity. Distances are measured by using so-called standard candles, objects (typically a type of star) that give off the same brightness and whose distances can be judged first in nearby cases, and then can be traced to the same kinds of stars at great distances. Hubble's successor at Mount Wilson, Alan Sandage, delved deeper into the universe with the aid of the two-hundred-inch telescope on Mount Palomar in Southern California. By applying Hubble's law farther and farther out, Sandage realized that distant galaxies had recession velocities of 10 percent or more of the speed of light.

The second proof of the big bang comes from the prediction that the early explosion of the universe would produce light elements with the simplest atomic structure. That would mean mostly hydrogen, but also helium and traces of deuterium and lithium. In fact, the universe indeed is abundant with these light elements. This prediction came largely from the insights of George Gamow, a Russian refugee to the United States in 1934. Gamow was a nuclear physicist with a remarkably broad perspective. He initiated our modern understanding of thermonuclear fusion. How could a pair of protons merge together, he asked, to eventually form helium? Resolving this paradox led to our understanding of how stars shine and to the development of the hydrogen bomb.

Eventually Gamow became interested in the big bang theory.

Until then, the prevailing view was that the big bang was a cold event that, nevertheless, led to expansion. But the cold theory had a problem. At the cold state, which means a very high density, atoms merge into the most tightly bound nucleus, which is iron. Hence, iron should be the most plentiful atom in the universe, which is certainly not the case.

Gamow realized that a moment of extreme heat could circumvent this problem. The first nuclear reactions, which produce new elements, could not begin until the nuclei of atoms had overcome the heat and then begun to merge with others. At the start of the universe, he reasoned, there was only a small space of time, perhaps only a few minutes, for the light elements so abundant in the universe to form before the cooling down produced the heavy elements. At some early instant, the universe had produced hydrogen, which today remains the predominant element. It must have been a hot beginning.

Following this logic, Gamow was the first to see that cosmology presented the ideal conditions to understand the origin of light elements. Of course, for the light elements to have seeded the many other nuclear reactions that followed, the universe must have been exceedingly hot. But no one had detected any residual heat in the universe, as predicted by a hot big bang, so the prevailing wisdom favored a cold origin. Decades later, the cold theory was still being advanced by the great Russian cosmologist Yaakov Berisovich Zel'dovich, and Gamow's contribution was mostly forgotten.

However, lack of evidence for a hot universe did not deter Gamow's genius for advocating a new way of doing cosmology. He now turned to stars. Gamow believed, erroneously as it turned out, that stars did not have the thermonuclear power necessary to synthesize the chemical elements seen in the universe. He was partly right. Indeed, stars cannot make a significant quantity of light elements, although they can make some. With stars eliminated, Gamow argued, the universe itself was the ideal place for synthesizing the second most abundant element, helium. He proved his prediction by describing how the universe expanded in phases, briefly achieving temperatures above those in the sun. A phase needed to last only a few minutes, he argued, to produce light elements.

This takes us back to the start of the universe. The only massive particles that could have lived on from the beginning were protons, neutrons, and electrons. Protons and electrons make up hydrogen, which is why that is the most abundant element today. Initially it was too hot for atoms to exist. The dominant constituent of the universe was ionized hydrogen, which meant it now had an electrical charge and was the only chemical element in the early universe. A few neutrons, about one for every proton, were present. For the universe to move toward formation of other chemical elements, a great deal of heat was needed so nuclear reactions could combine protons and neutrons. With the heat, protons could overcome the force of Coulomb repulsion (that keeps particles apart) and form elements heavier than hydrogen. Gamow had discovered this natural barrier. With his insights, we find the beginning of nuclear physics as a new branch of science. His brilliant idea was that if the universe were, by fiat, initially hot, the required nuclear reactions would have taken place in the first minutes.

He enlisted his student Ralph Alpher and his colleague Robert Herman into the research project, which culminated in predictions of the exact abundance of light elements in the universe. Helium, amounting to some 30 percent of the mass in the universe, was synthesized in the first minutes of the universe. Previously a mystery, the origin of helium—the second most abundant element—was now resolved.

Gamow's dream of the origin of all of the chemical elements had one problem, however. Only 2 percent of the elements produced in his hot big bang were heavier elements, such as traces of lithium and beryllium. Where did all of the heavy elements come from? As we now know, the heavier elements are made in exploding stars called supernovae, which scatter their ashes around the galaxy. Nevertheless, Gamow was fond of joking that his theory should be considered a success, and rightly so. It explained the nature of 98 percent of the matter in the universe: hydrogen and helium. However, the most important proposal of his theory, the expectation of a hot universe, was to remain forgotten for nearly two decades.

That changed in 1964, when the background radiation of a hot big bang, once predicted, was now discovered quite by accident, resulting in our third piece of evidence for the big bang. In New Jersey, the radio astronomers Arno Penzias and Robert Wilson had gained access to a microwave radio telescope. It was originally designed for the first satellite communications system, but was subsequently overtaken by better technology. Using this old equipment, Penzias and Wilson wanted to survey only the Milky Way galaxy. But they found beyond the Milky Way a pervasive glow that is apparently isotropic, that is, has the same structure in all directions. (Being isotropic is not necessarily the same as being homogeneous, however.) Furthermore, the glow seemed to have no relation to our galaxy. They at first refused to believe their measurements. They tried to explain it away as an experimental artifact but they did not succeed.

Then a bit of news put them into action. When they heard that a rival group at Princeton was searching for the fossil glow of the big bang, Penzias and Wilson realized what they had discovered. They promptly published their measurement of excess radiation—the beginning of a long process of proving Gamow's hot big bang theory. Indeed, Penzias and Wilson were initially unaware of his arguments. Nevertheless, they had found the elusive background from the beginning of the universe.

At last, Gamow and his collaborators were vindicated. However, they never received full recognition for their prediction of an initially hot universe that produced relic radiation. They did not really appreciate the need to connect their theory with microwave astronomy. In their publications as well, they did not use terminology that would have caught the attention of microwave astronomers, such as Penzias and Wilson. I recall once encountering George Gamow surrounded by a small crowd of astronomers, as he declaimied in his high-pitched voice that he "had lost a penny, Penzias and Wilson had found a penny, and was it his penny?"

After the serendipitous discovery by Penzias and Wilson, much of the early debate about the hot big bang came to a climax in early 1967. The chief forum turned out to be a cosmology conference at the Goddard Institute for Space Studies in New York, where the topic was intensely debated. For historical reasons, this meeting was called the Third Texas Symposium on Relativistic Astrophysics, following earlier meetings in a series at Dallas and Austin. These were heady days in astronomy and cosmology. At the first Texas meeting in 1963, the superstars were quasars, just discovered (although their true nature and distance still are debated). The newly named idea of a black hole, a singularity in space of nearly absolute density, much like the start of the universe, was also announced at one of the Texas symposia.

The 1967 New York meeting marked a turning point for acceptance of the big bang theory. Before then, the very name was a kind of slur. It was coined pejoratively by British cosmologist Fred Hoyle, who favored the rival steady state model of the universe. As Hoyle famously told a popular BBC radio broadcast in 1950, the idea of "big bang" was "an irrational process that cannot be described in scientific terms . . . [or] challenged by an appeal to observation."

Even after the discovery made by Penzias and Wilson, however, it took another fifteen years to verify the exact nature of this cosmic background radiation. According to predictions, it had to be what we call blackbody radiation. A corollary of the light element interpretation was the prediction that the cosmic microwave background would show a blackbody spectrum, which has so completely mixed the different wavelengths of heat that it approximates the evenness of a perfect furnace. The first prediction of this perfect blackbody temperature was made by Alpher and Herman. But applying such calculations to microwave astronomy did not come for another two decades, when Robert Dicke at Princeton (the rival whom Penzias and Wilson had heard about) was arriving at the best estimates of the background radiation temperature.

The blackbody spectrum would be expected if it emerged from a dense and hot beginning. The principal attribute of a blackbody spectrum is that it carries no information about the sources of heat and radiation. Everything is mixed, where just as in the ultimate car wreck, a Ferrari cannot be distinguished from a Ford. This is a state of maximum entropy, or disorder, and hence we have the absolute minimum of information. According to the second law of thermodynamics, a fundamental law of physics, the order of a dynamic system, left on its own, can only remain constant or decrease. Chaos reigns. But as structure develops, the entropy is decreased. Curiously, as stars and planets form, creating structure, they also radiate heat, which is like expelling the entropy from the system. This is exactly what happened in the formation of the Milky Way galaxy, for example. The entropy of the universe as a whole is conserved, but at the same time, we see the rise of structures of increasing complexity, such as galaxies, stars, planets, and life itself.

The entropy of the universe is measured by its radiation content. We know this because the number of photons in the cosmic microwave background vastly outnumbers the number of particles in the universe. This is true for any high-entropy system. Blackbody photons are the ultimate cosmic equalizer in terms of lack of any preferred information content.

The first definitive spectral measurement of the microwave background was made by the Cosmic Background Explorer (COBE) satellite. The satellite carried sensitive experiments that were designed to measure the frequency distribution of the sky photons as well as tiny differences in the sky temperature to a level better than anything previously accomplished. This accomplishment did not come quickly. It took seventeen years to launch the satellite after it was first conceived. Long delays are common in space astronomy, but this one was exceptional. The delay of COBE was compounded by the *Challenger* space shuttle disaster in 1986. With the curtailment of space shuttle missions, NASA had to reconfigure the COBE satellite for launch on a Delta rocket, which took place finally in 1989.

One experiment on the COBE satellite measured the spectral energy distribution of the cosmic background photons, which is a clue to their origin. The Penzias and Wilson measurements suggested it would lie almost entirely in the far infrared and microwave frequency region. This very red distribution corresponds to extremely cold radiation, at only 3 degrees Kelvin. Compared to early instruments, however, COBE's measurement of the spectrum was remarkable for its precision.[1] The temperature was measured to be 2.736 degrees Kelvin. It is a perfect blackbody. In fact, a better blackbody spectrum cannot be synthesized on earth. The universe emerged from an ideal furnace.

A few seconds after the start of the big bang, the temperature was about a billion degrees, but the universe began to cool quickly. The blackbody spectrum found by COBE corresponds to what the big bang would have been like for the first few weeks, a perfect furnace. During this phase, the universe was completely opaque. Matter was in the form of hot plasma. The plasma was mostly ionized hydrogen, produced when the electrons cooled enough to be captured by protons. Soon hydrogen atoms became the dominant constituent of the universe, at least in terms of ordinary matter. Hydrogen is ineffective at scattering the radiation, and so about three hundred thousand years after the big bang, the universe became transparent (at a cool temperature of about 3,000 degrees Kelvin). We can see freely back to this point in time. As we measure the radiation today, we are seeing back to that early phase of the universe, just by looking in a dark patch of the sky.

1. This was the Far Infrared Infrared Spectrometer, designed by a team led by John Mather, who shared the 2006 Nobel Prize in physics with George Smoot.

The Final Piece

We now come to the final line of evidence that has been crucial in verifying the big bang: confirmation of the fluctuations in the cosmic microwave background. Without these, the entire edifice of the big bang might have collapsed like a house of cards. We needed to explain the original seeds of the structure we see in the universe today, and after a long search these seeds or fluctuations have been found.

The fluctuations must have arisen at an early phase of the universe, when matter no longer was in competition (by friction) with the radiation. The matter would naturally be slightly denser in some regions, compared to the average background, and thus have the gravity to attract more matter. This slight overdensity increases. This movement is not in one direction, however, because every overdensity is matched by an underdensity. The underdense regions become emptier. The overdensities are the precursors of galaxies. This early back-and-forth of density produced a pattern of fluctuations that should, as astronomers predicted, show up in the background radiation.

The physics of density fluctuations has proved to be remarkably fruitful in all areas of cosmology, from the beginning of the universe to its large-scale structure and the formation of galaxies. It all begins with the very nature of gravity. Gravitation is intrinsically attractive. Pressure opposes gravity, but in a large enough system, gravity is overwhelming. It results in fragmentation. We refer to this process as gravitational instability. Overdensities grow at the expense of underdense regions. By analogy with capitalism, the rich get richer. When we look at the universe today, we see structures, which tells us that there must have been gravitational instability at the very start of the expanding universe. In other words, the universe could not have been completely uniform when the expansion began. If it were uniform, structures such as our Milky Way galaxy would never have developed. Finite density fluctuations must have been present from the beginning in order for structure to have evolved. But the predicted value of the seed fluctuations was a fraction of a percent in the very early universe.

Once the cosmic microwave background was discovered, beginning in 1964, cosmologists finally had a backdrop against which to measure accurately the fluctuations of the early universe. These infinitesimal variations are the only way we have to probe the initial conditions of the universe. In principle at least, the fluctuations would show up in tiny differences in temperature. The fluctuations, as ancient relics, would reflect the kind of large-scale structure and dispersion of galaxies we see in the universe today. In other words, today we see the relics of the big bang in only three forms: one being the light elements, the next and complementary form being the cosmic microwave background, and the third being the structure of matter. We call the heavy particles that produced the first matter baryons, a general term for the protons, neutrons, and electrons that make up every element.[2] Baryons are relics of the early universe. They are the stuff that constitutes our bodies today.

Once the idea of fluctuations was clear in cosmology, the race began to find them in the sky. Over the past decade, these measurements have become increasingly accurate. When it comes to predictions on the kind of fluctuations that seeded the early universe, in fact, no significant inconsistencies have been found. In all, this has been a final dramatic verification of the big bang theory.

The research on fluctuations has given us a great insight into the formation of galaxies. In confirmation of the big bang, we have seen that the distribution of galaxies shows us what the fluctuations must have been like back at the beginning. We observe a highly uneven distribution of galaxies, so we can assume that such uneven irregularities must have been present before the galaxies formed. Now we move forward and imagine how this process has taken place.

2. Light elements are also baryons, in particular helium, which constitutes one third of all the baryonic matter.

Imagine a region that is large enough to contain all of the mass that now makes up a galaxy, but is very slightly overdense, by a fraction of a percent. In fact, this seems small but is nevertheless highly improbable if left to chance. As noted earlier, density fluctuations are able to grow in strength. The slightest excess in gravity attracts surrounding matter. A small overdensity eventually becomes larger. This region of density still expands along with the universe, but at a slower rate, so it lags behind. When the contrast with its surroundings is large enough, the region becomes dominated by its own self-gravity. This region is cooling down, which means its outward pressure is being overcome by the gravity of its matter. At this point, the region is gravitationally unstable. The region must collapse to form a massive cloud, which becomes the birthplace of future galaxies.

Inside the cloud, the density fluctuations also cause it to begin breaking up into clumps. Even in a static cloud, infinitesimally small fluctuations could grow exponentially. A slight excess in density would attract matter from its vicinity. The fluctuation gains strength. But now, the expansion of the universe means that any region about to collapse into a cloud is itself expanding. It expands until its own self-gravity becomes dominant. The expansion quenches the rapid growth, but slow growth still occurs. Overdense regions continue to become denser by drawing matter from the surroundings, but more gradually than in the case of a static cloud.

The expansion guarantees that it takes a long time for the overdense region to form a self-gravitating cloud. It is as though one is running on a continuously expanding track. It takes much longer to finish the race. Nevertheless, at the start of the universe, such clouds are destined to form the first galaxies. The typical masses of galaxies are initially small, only a few tens of millions of solar masses. In time, however, larger and larger clouds condense, and after a billion years most galaxies are in place. The universe is then one fifth of its present size.

Galaxies continued to form over the next 10 billion years. As mentioned earlier, we can tell the younger and older galaxies by their shapes and colors. The older galaxies have elliptical shapes with predominantly older stars. The younger galaxies have spiral structures. They are disk-shaped and often have a central spheroidal bulge of old stars at the center. Galaxy colors also tell us the ages, and colors depend on the mix of stars. A blue star is massive, hot, and short-lived. A red star has low mass, is cool, and is long-lived. We find that spiral galaxies are blue whereas elliptical galaxies are red. In fact, the old red stars are the dominant constituent of most galaxies, having formed when the universe was young, and also forming the spheroidal component at the center of galaxies. Disks are intrinsically blue, although partly reddened by interstellar dust and by the underlying component of older stars.

Today, star formation continues in the gas-rich disks of galaxies, and especially in the spiral arms. Ongoing star formation traces out beautiful spiral patterns. Our sun, for example, is at the edge of the Orion spiral arm in the Milky Way galaxy. The nearby Orion nebula demarcates the most active star-forming region in our neighborhood. It is a nursery for star formation.

We have come a long way: from the big bang furnace through the early fluctuations to the large-scale structure of the universe. The one thing that was not anticipated, or even understood until very recently, is that the expansion of the universe is accelerating. Recall that in Einstein's theory of general relativity, gravity causes deceleration. This is a positive pressure that enhances gravity, moving the universe toward collapse. But we find that the universe is not only expanding, but it is also speeding up. Something is providing the negative pressure that counters gravity, causing the acceleration ands leading to one of our most baffling theories in modern cosmology—a negative pressure, or cosmological constant, that we call dark energy for lack of a better term.

Overall, we have just covered what is now the standard model of cosmology. It brings us to one more remarkable conclusion about the universe. It has reached the critical density that is required for the universe to be structured according to Euclidean geometry, meaning we can measure the universe as if it were flat. But meanwhile, much of this flat universe is made up of something we cannot see, namely dark energy and dark matter, Two-thirds of the mass-energy density of the universe is made up of dark energy, according to our calculations. We don't know what dark matter is made of, but we know that it makes up the other third of the density of the universe. The mystery only deepens, for just 15 percent of the dark matter is composed of baryons, the stuff that forms stars and planets. As a later chapter shows, cosmology is deeply challenged by the quest to decipher the exact nature of dark matter, or to prove whether it even truly exists.

Particle Physics and Inflation

Up to now, the great theorists of our universe have been astronomers and cosmologists, people such as Friedmann, Lemaître, Einstein, De Sitter, and later the British astronomer Arthur Eddington. They are immortalized for various models of the expanding universe, at least before we arrived at the conventional model of today. What they had always lacked, however, was a theory about the initial conditions of the universe, in the veritable first moments of the hot big bang. The cosmology coming out of the 1930s did not have a theory. A new idea was needed. It finally came from a group of scientists who hailed from outside of astronomy, the particle physicists.

As outsiders, their arrival jarred astronomy but also was a new impetus for advances in cosmology. Particle physics was the field of nuclear physics that pioneered the understanding of the origins of helium, for example. The particle physicists looked at the big bang as a great particle accelerator that had achieved the highest energies possible in the universe. It had conducted one great experiment long ago, and we can now look at the consequences. To do particle physics, they had to think this way. On earth, it is hard to imagine constructing any particle accelerator more than a hundred kilometers across. The size of such technology is limited, and so are budgets. For a machine to match the energies at the first instants of the big bang, for example, we would need an array of superconducting magnets that stretch to the moon or beyond. The benefits of focusing on the beginning of the universe became obvious. One can study the effects of ultrahigh energy particle interactions for free, or, at least, for a relatively trivial sum.

This led to the field of inflationary cosmology, which first emerged in the 1980s by asking the question, why is the universe, despite its likely irregular structure at the beginning, now overall fairly smooth, even, and flat in the Euclidian sense? The answer seemed to be that at the start of the universe, there might have been a rapid inflation of matter, evenly mixing up its forces and particles before it began its path of expansion.

How did they arrive at an inflationary theory? It starts with basic physics. At very high energies, some new physics comes into play. High-energy physics uses the analogy of symmetry. All fundamental forces have the same strength. However, as the universe cools and expands, this symmetry is broken. We can imagine symmetry by thinking what would happen to a Rolls-Royce and Volkswagen as each is exposed to extreme heat. At first they remain distinctly different. Soon, they would melt into a soup of inorganic chemicals. No doubt there would still be differences. But at even higher heat, one would be left with the constituent chemical elements. The cars would be indistinguishable. Symmetry prevails. This was the beginning of the universe, paradise for some but closer to hell for others.

We can also go in the other direction, which is the breaking of symmetry. This is the story of the formation of galaxies. As the temperature of the universe drops, chemical and physical forces separate because the force that holds the atomic nuclei together is stronger, and on a smaller scale, than the force that holds atoms together. The cooling also allows the electrons that are now orbiting the nucleus, like planets around our sun, to begin a process of electromagnetic interactions, which control chemistry. This allows hydrogen to form. The cooling universe is on its way to forming stars.

All of this has happened because the original symmetry between the fundamental forces of the universe broke down. Let's start at the very beginning of that process. At the highest energies there were no nuclei, not even protons or neutrons. The natural state of matter was a soup of quarks and gluons. A proton is made up of quarks that are held (or "glued") together by clouds of gluons. We are in the realm of quark-gluon plasma, the dominant state of the universe about a tenth of a nanosecond after the big bang. The phase transitions continue, each a change in the state of matter that is accompanied by a release of energy. It is much like the energy given off when ice melts into water, a global warming if you will.

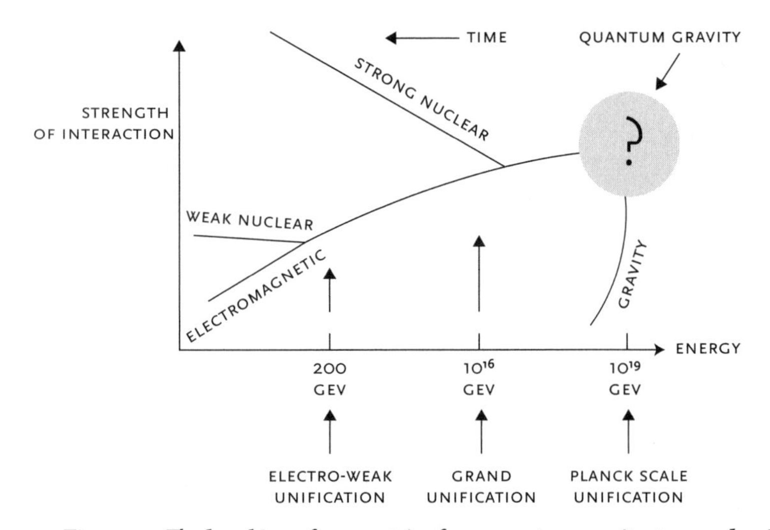

Figure 2.1. The breaking of symmetries from quantum gravity, to grand unification, and to the appearance of our low energy universe when the electromagnetic and nuclear forces separate.

The next instant is what we call the era of electroweak unification, when the electromagnetic and weak nuclear forces of the uni-

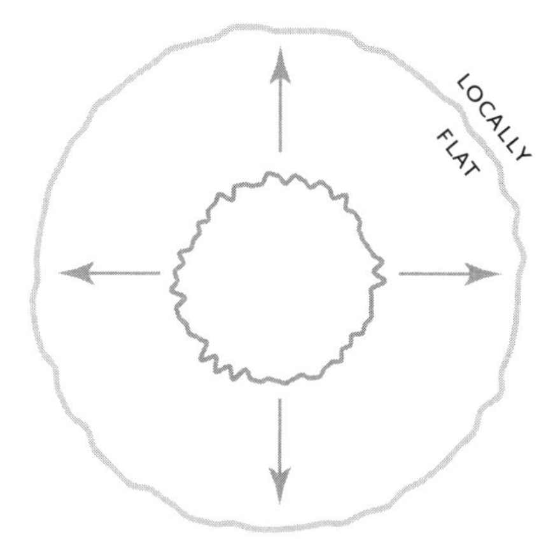

Figure 2.2. How inflation removes all initial irregularities in the universe and simultaneously explains why it is so large and spatially flat.

verse were in total unity. This moment comes an instant after the big bang, a fraction of time so small that it has fifteen zeros after the decimal point: 0.0000000000000001 of a second (one-quadrillionth of a second). This was not a complete unification, however, for the strong nuclear force had already broken away from the party. We call this earlier moment the epoch of grand unification when electromagnetic, weak and strong, nuclear forces were all indentical. Gravity was united with the other forces even earlier, but that was at a still higher temperature, which we call the Planck instant, also after German physicist Max Planck. This realm of Planck physics is poorly understood, so for the moment, we bypass this very beginning point of symmetry.

Returning now to the moment of grand unification, it too breaks down as the temperature drops. More symmetry is lost. As mentioned earlier, the strong nuclear force becomes stronger than the weak force. They break apart. The weak force remains indistinguishable from the electromagnetic force until much more cooling occurs. Then the weak force separates. The electromagnetic force is also free to operate, based on the electrons that orbit around a nucleus of the atom. Chemistry has begun. Its first product, much later, is hydrogen. So many phase transitions are happening at this point in time that a very substantial amount of energy is injected into the universe. One of these releases was so powerful that it inflated the size of the universe in an instant.

The heart of inflationary theory is that the inflationary release of energy does not last long, but it manages to provide a big transient boost to the energy content of the universe. The energy release counters the effect of gravity in a dramatic way. The effect produces repulsive gravitational forces, which act like antigravity. The expansion rate of the universe is boosted. There is so much antigravity that exponential expansion sets in. Hence, we call this inflation—indeed, the mother of all inflations.

We can imagine this transformation taking place in an infinitesimal patch of the universe. It can be anywhere, but it seems inevitable somewhere. Once such a patch develops, it switches from having a strong gravitational attraction to having a strong gravitational repulsion. Powered by this phase transition energy, the patch grows exponentially, inflating the universe just as suddenly. In fact, this infinitesimal patch has now encompassed almost the entire universe. By virtue of this rapid and immense expansion, the universe becomes almost completely uniform.

This answers the baffling mystery of the universe's general uniformity. With the inflation there is an associated flattening of the geometry. The geometry of space may once have been non-Euclidean. Perhaps all space was curved before the moment of inflation. But after an immense expansion, space flattens out, at least locally. This is what we mean by an ironing out of possible wrinkles in space-time. Space is wrinkle-free, and it is much, much larger than the observable universe today. Inflation gives us a reason. The universe may once have been a messy environment. It could have expanded more rapidly in one direction, leading to a cigarlike shape of space. Or in two directions, resulting in a giant pancake. However, inflation smoothed away nearly all such deviations from uniformity and sphericity.

All of this occurred very soon after the big bang. The way we measure this event is by thinking of about a trillionth of a trillionth of a trillionth of a second—a decimal point followed by thirty-five zeros, or $.10^{-36}$ seconds. Let's call this a googolth of a second.[3] Once ten googolths of a second have elapsed, inflation is over. The phase transition is complete and inflation stops. The universe is now very large but very cold. Fortunately lots of energy is left over from the inflationary epoch, which causes the universe to reheat, and the usual hot big bang resumes as the universe continues to expand.

We have gone back nearly to the beginning, but there is a barrier in time that we cannot penetrate. This is even earlier, about a ten-millionth of a googolth of a second after the big bang, the sought-after instant when gravity was joined with all the other forces. At this moment, the strength of the gravitational force equaled that of the nuclear forces. This is the Planck instant.[4]

We have no theory that describes this state, which is where quantum theory meets gravity. Arriving at the theory of quantum gravity, the ultimate unity of forces, is the ultimate goal of modern physics. Perhaps this theory will provide the explanation of why inflation occurred. Right now, inflation is a very good conjecture, one that has predictable and verifiable consequences. For us, the most important of these is the emergence of structure.

In 1980, particle physicists such as Andrei Linde, then in Moscow, and Alan Guth at the Massachusetts Institute of Technology provided us with some of our most innovative theories of how inflation took place. They were outsiders to astronomy, but they too realized that the big bang provided a natural particle accelerator. It had achieved energies vastly higher than any man-built machine could ever accomplish. For example, the energy at the big bang was 10^{16} GeV. We can contrast this with the largest terrestrial machine

3. Actually a googol is 1 followed by a hundred zeros. Google founders Larry Page and Sergey Brin based their company's name on "googol."

4. The Planck instant is 10^{-43} seconds.

we have, the Large Hadron Collider, a new twenty-six-kilometer (seventeen-mile) circumference particle accelerator straddling the Franco-Swiss frontier. It began operation in 2009 under the supervision of the European Organization of Nuclear Research (CERN). As big as it is, however, the accelerator generates just 1,000 GeV, less than the big bang by a factor of more than a thousand trillion.

The Large Hadron Collider is likely to move us closer to an understanding of the early universe. But as a result of the energy discrepancy, our prospects of verifying grand unification and the phase transition that triggers inflation are remote. In the meantime, the best we can hope to do is to pin down the outcome of inflation. At least two distinct signatures can help us distinguish between competing models of inflation. Both come from the inflationary epoch.

One is a background of gravity waves. These are produced at the end of inflation along with the density fluctuations that seed large-scale structure. The difference is that gravity waves do not seed anything. No compression is involved as a gravity wave passes by. The wave only has a shearing motion. Still, these waves are a signature of inflation. The difficulty in detecting them is that they are highly red shifted. The wave frequency observable today is a thousandth of a hertz or less. Terrestrial backgrounds interfere at such low frequencies. These include the ocean tides. To detect such low frequencies, one has to go into space and use satellites that are millions of kilometers apart. Laser beams connect the satellites. As a gravity wave passes, one has to measure a change in the length of the laser beam of a centimeter or less to arrive at the expected frequency range where a signal might be detectable. In 2020, such an experiment, LISA, is jointly scheduled for launch by the National Aeronautics and Space Administration (NASA) and the European Space Agency (ESA).

Another signature of the original inflation is the distribution of fluctuations in the cosmic microwave background on angular scales more than a degree. New experiments will study these fluctuations and look for the inflationary imprint of gravitational waves on the fluctuations. One effect is a slight polarization. To detect this effect, a sensitivity of a hundredth of a millionth of a degree Kelvin is required, a hundred or more times better than our currently achievable sensitivities. Future space experiments will be needed to achieve this goal.

The search for inflation has taken us into space. It has also forced us to try to explain the large-scale structures of the universe, which is the topic of the next chapter.

CHAPTER 3
Inflation Explained

COSMOLOGY PICTURES a time when tiny fluctuations, generated during an instant of inflation, produced all that we see in our vast universe. The inflation of those tiny irregularities determined the large-scale structures of the universe billions of years later. This is quite a picture. It means that a slight overdensity of energy at the beginning of time has brought about the formation of our magnificent galaxies. That is a link we pursue in this chapter.

As mentioned earlier, the way gravity operates is such that "the rich grow richer." Capitalism prevails, at least in the origins of structure. A contrast in density makes the density grow. But how did this contrast begin? This is a question of the very earliest fluctuations. We have to turn to the quantum theory of the atom. Quantum theory tells us that the most perfect vacuum is not really empty. At any place, at any moment, a quantum fluctuation can arise in the energy of the vacuum. This manifests itself as a slight excess or deficit in density. As the universe inflated, there was ample time for the quantum fluctuations to be stretched and amplified onto macroscopic scales. One remarkable aspect of this concept is that an emergent democracy arises from capitalistic beginnings. Whether the mass-content of the fluctuation is the size of a pea or a galaxy, its emergent strength is the same.

What is still unclear in inflation theory is the strength of the original fluctuation itself. This hole in the theory has motivated a wide array of variations on the early inflation model. The simplest models are characterized by having only one inflating force field. The more complicated variants have multiple fields. Today, cosmologists have a hard time selecting a preferred theory of inflation, and the wide variety makes it very difficult to make predictions that can be tested.

Nearly any astronomical observations can match an inflation theory that is sufficiently tweaked. In the 1990s, astronomical data seemed to favor an open universe, one whose geometry resembles what one would find if confined to the surface of a hyperboloid. Hence, inflation theory was tweaked to allow for an open universe.

Now our best data prefers a spatially flat universe, and we can expect inflation theory to adapt to these findings as well. A flat universe is one whose geometry is that of an infinite flat surface or plane. Euclid's axioms are satisfied on the plane, meaning that the three angles of a triangle add up to 180 degrees. Our astronomical observations favor Euclidean geometry, and now most inflation models result in a flat universe. There is a rival, of course, and that is a group of more convoluted models of inflation that predict curved space. These seem very unlikely, however. So we can say that a key prediction of inflation—the flatness of space—is more or less verified.

Admittedly, we are voting by majority. This is often how conventional theory develops in science. We can understand this human factor by the use of statistics. Even today, we rely on a statistical argument laid down in the eighteenth century by the British Presbyterian clergyman Thomas Bayes (d. 1761), now called Bayes' theorem. He argued that the probability of an outcome is dominated by the number of possibilities, in our case, the number of models of inflation. Most of our models favor flatness. Hence, we infer that flatness is a likely outcome of inflation.

Despite this apparent adoption of theory by majority vote, inflationary cosmology in general has an elegance that provides compelling explanations for a number of observed phenomena which include more than just the observed near flatness of the universe.

There is also the universe's vast size and near homogeneity. Inflation seems to explain the small deviations from homogeneity that are so important for understanding the origin of the structure of the universe.

To understand inflation, cosmologists have tried to measure the fluctuations in the universe. Their prediction has been that these would correlate with the past, and with the expansion of the universe, which is exactly what they have found. To test this idea, imagine measuring the typical level of fluctuations in a series of ever-increasing regions of the expanding universe. This step is made possible by our deep galaxy surveys with large telescopes. We find the following: The larger the region, the more the universe approaches homogeneity. On average, the universe is completely homogeneous. There is no dense center, no rarified boundary region. Yet everywhere there are galaxies. In some regions, there are slightly more than the average, and in others, slightly fewer. We describe these variations as fluctuations in the average density of the universe. Some are positive, some are negative.

When we measure the strength of the density fluctuations, in other words, we find that the overdensity or underdensity is smaller with increasing scale. In fact, overdensity and size are found to oppose each other if we consider the strength of the gravity field. All fluctuations are found to have the same strength if measured by the strength of the associated gravitational potential. We observe strength independence for the distribution of density fluctuations in the large-scale galaxy distribution. Again, this is predicted by the simplest inflation theories.

An important concept for studying inflation is the horizon, which refers to the edge of the universe as it expands from the original big bang. During the expansion, the horizon is the distance traversed by a ray of light since the big bang. With time, the horizon grows outward as the universe become larger. In practical terms, the horizon is the limit of human perception, by our astronomical technology, of the edge of the expanding universe. Now imagine the beginning. Inflation begins when the horizon contains the quantum equivalent of a handful of atoms. Actually, it is more than a handful, rather what we refer to as a Planck mass of about 0.000001 gram. This is the scale at which quantum theory meets general relativity.

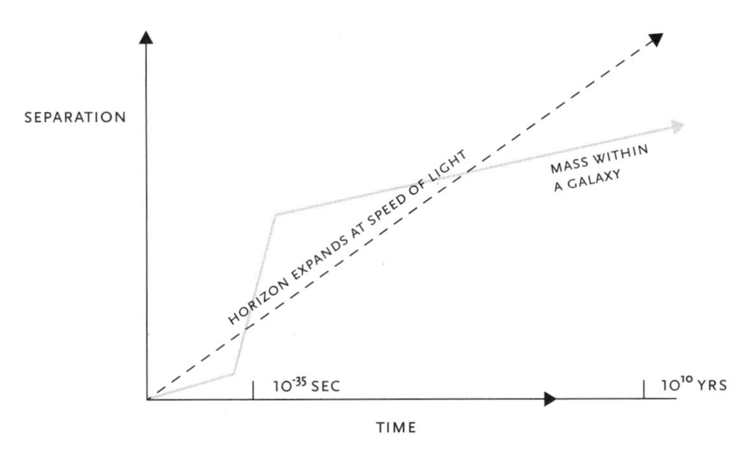

Figure 3.1. The horizon of the universe, compared to the physical scale of the mass within a galaxy. Time advances from the beginning until the present.

The horizon, were it any smaller, would contain even denser matter and would inevitably form a black hole (a total collapse of density). Nothing can escape from a black hole, not even the expanding universe. This contradiction is another way of telling us that known theory breaks down at this point, some 10^{-43} second after the big bang. We cannot extrapolate further backward in time with any confidence. But it is at this moment that we can be sure that fluctuations are large. In a sense, this is a macroscopic manifestation of quantum uncertainty. Inflation sets in somewhat later, when there are still significant density fluctuations present. It grabs these fluctuations and transports them to the vastly larger scales over which, much later, galaxies will form.

Inflation-generated fluctuations provide the density fluctuations that eventually, billions of years later, seed galaxy formation. These fluctuations are frozen for the first ten thousand years of the big bang. Only when the density of the universe has more matter than radiation does any change occur. From this moment on, the fluctuations grow in strength. A slight overdensity pulls in surrounding matter by gravitational attraction. It becomes relatively denser. A slight underdensity evacuates matter. It becomes relatively emptier. We call this process gravitational instability. This instability, however, is stronger in regions close together than those expanding farther out. Regions that can communicate with each other grow by that causal connection. The growth only occurs on scales smaller than the horizon. Consequently, the smaller scales are advantaged over larger scales. The larger scales entered the horizon later and are farther out. Hence, the smaller scales acquire a head start in the race to form structures.

However, there is an obstacle to growth on all scales. This delay lasts for the first hundred thousand years of the universe, because the universe early on is mostly radiation. Growth comes only after the radiation density falls below the matter density. Meanwhile, the self-gravity of matter cannot be effective in a fluid of energy with such high pressure. The only growth that occurs is driven by the relatively cold dark matter in the universe (a topic to which we return). For the first ten thousand years, the universe is radiation-dominated, curtailing growth on subhorizon scales, but growth finally resumes in the matter-dominated era.

This difference between smaller and larger scales, in which small regions form galaxies first, is called bottom-up evolution of galaxies and it works in this way. When matter does dominate in the expanding, cooling universe, the horizon contains approximately the mass equivalent of a great cluster of galaxies. On smaller scales, all earlier growth stagnated. As larger and larger scales enter the horizon, they begin to grow. The subcluster scales (which are smaller than the first clusters) undergo maximum growth. Because of growth stagnation on scales below the horizon size at the epoch of matter-radiation equality, the resulting distribution of fluctuations has the strongest fluctuations on the smaller subcluster scales.

On larger scales, growth of fluctuations occurs systematically later as the horizon grows, so that by today the initial fluctuation amplitude after the big bang is seen only on the largest visible scale, the current epoch horizon scale. Smaller scales, which entered the horizon earlier, grow more than larger scales. On the smallest scales, well below the horizon scale at matter-radiation equality, all fluctuations grow throughout matter domination and so undergo a similar amount of growth. The net effect is a distribution of seed density fluctuations with larger amplitudes on smaller scales. The fluctuations continue to grow by gravitational instability in the matter-dominated epoch. Eventually galaxies form. Small galaxies are destined to form before their more massive counterparts—hence, bottom-up evolution.

Dark matter, according to the theory of how galaxies formed, must enter the picture here. It is called dark matter because we cannot see it and do not know its nature. It nevertheless shows up as a major gravitational force in the universe. One of its main characteristics is that it has a weak interaction with other particles, including heat, and we know that it was somehow present in the early expansion of the universe. By having gravity, dark matter helped boost the growth of the early fluctuations. It also aids in galaxy formation.

In the radiation-dominated era, before matter dominated, we have a variety of forces working together on the different particles. There is a competition between radiation pressure and gravity. These forces act on the fluctuations, which we can think of as sound waves in a fluid consisting mostly of photons of light. The random motions in a gas of photons are near the speed of light. In fact, the pressure in the fluctuations drives waves traveling at the sound speed, which is around 60 percent of the speed of light.

Two other particles are present, the baryons (heavy particles) and the weakly interacting dark matter particles. They will move in different ways as the universe begins to cool. The baryons are destined to form dense clouds. The nonbaryonic dark matter has similar fluctuations to the baryons, but does not interact with the radiation. Therefore, as the universe becomes matter-dominated, the dark matter fluctuations grow in strength. In contrast, the baryons are still coupled to the photons and continue to oscillate as sound waves. Now the dark matter fluctuations have their own distinct signature. They are destined to form the dark halos of galaxies. Halos are vast clouds of dark matter within which are embedded in the galaxies that we observe. Every galaxy has its own dark halo.

As mentioned, dark matter boosted fluctuations in the universe. That process can be explained by going back to when the universe was a plasma. The plasma consists of free electrons and protons, with a 10 percent sprinkling of helium nuclei. The photons are scattered in all directions. But then something dramatic happens when the radiation temperature drops below three thousand degrees Kelvin. There are no longer enough energetic photons around to keep the matter ionized. The hydrogen becomes predominantly atomic. Very few free electrons are left behind. The photons no longer are scattered, and the radiation is decoupled from the matter. The baryons fall into the gravity potential wells of the dark matter fluctuations. In this way, the growth of fluctuations is boosted due to the presence of dark matter.

All of this activity is driven by gravitational instability, and that process continues. Indeed there is no longer any braking on the gravitational instability of matter that was once provided by friction with photons. So structure develops on the smallest scales in both the dark matter and in the baryons. Eventually, the fluctuations are dense enough to separate from the average density field. They are no longer fluctuations, but rather assemblies of self-gravitating dark matter and baryons. Larger and larger systems condense out as the universe expands. The dark halos of galaxies have formed.

Once dark matter clumps form that are sufficiently massive for the baryons to be able to cool, galaxy and star formation commence in earnest. Cooling is a necessary condition for star formation. The gas must be hot enough for collisions between atoms to cause atoms to radiate. This requires a cloud temperature of at least a thousand degrees Kelvin. The smallest clouds that can cool must have masses of at least a million solar masses. The lowest mass clouds have less gravity and are cooler. They also condense out before their more massive counterparts. These clouds cool most effectively and become dense enough to form the first stars.

As time proceeds, the smallest clouds aggregate and accrete into larger clouds. The formation sequence is bottom-up, with smaller clouds forming before larger clouds, and with occasional stars now forming. Star formation remains sparse because it only takes the first massive star to blow its parent cloud apart. And as we shall see, the first stars are mostly massive stars that are destined to explode as supernovae. Eventually, we recognize the emerging clouds as galaxies.

The strength of the fluctuations determines when they will collapse into galaxies. The distribution of fluctuation strengths determines the epoch of galaxy formation. This distribution is set observationally by the strength of the observed fluctuations in the cosmic microwave background radiation. The first small galaxies formed when the universe was thirty times smaller than it is today. Hence, we say this occurred at a redshift of 30. In that epoch of galaxy formation, the universe was thirty thousand times denser than it is today.

As the universe expands, larger and larger galaxies form. Eventually, by a redshift of 5 (when the universe is one hundred times denser than today), massive galaxies are forming, the typical precursors of today's galaxies. These galaxies are initially just vast clouds of dark matter, called galaxy halos. About 10 percent of the mass in a halo is in the form of baryons. The principal difference between baryons and dark matter is that the baryons undergo nuclear and electromagnetic interactions. These interactions play a crucial role during galaxy formation. Atoms and molecules of hydrogen form. Collisions with these atoms absorb energy by exciting atomic or molecular energy levels. These excited atoms or molecules are deexcited as they radiate photons, letting the energy escape. Escape of energy is the essential requirement for a cloud to collapse.

This released energy of the baryons, called kinetic energy, is converted into radiation that escapes freely. The baryon cloud contracts, while the dark matter cloud in which it is embedded remains extended and diffuse at its initial low density. Dark matter cannot lose energy by radiation, so it cannot contract. The net result is that the baryons condense into dense clouds embedded in the dark halos. The clouds retain much of their initial angular momentum, the same force that we see at work when a top is able to stay upright as it spins. As the baryon gas loses energy and contracts, it spins more rapidly. Gaseous disks form, supported by their rotation. These disks are themselves gravitationally unstable, and break up into gas clouds which in turn fragment into stars. The epoch of galaxy formation has begun, a billion years after the big bang.

Detecting the Elusive Fluctuations

Once the reality of the cosmic background radiation was realized, with the 1964 discovery by Penzias and Wilson, a new quest began, to find and measure the fluctuations in the radiation and then use this finding to build a precise theory of galaxy formation. This is the story of theories, debates, and testing of predictions, a process of theorizing that actually stretched from 1946 into the 1990s, and included long periods of detective work.

The fundamental barrier to our understanding of galaxy formation was a lack of knowledge about the initial conditions of the early universe. Without this, astronomers were literally grasping in the dark. Because the early big bang was full of radiation, gravitational instability was almost totally minimized for the first ten thousand years. As noted, structures cannot grow until the universe has cooled down. At some point, we need to find fluctuations of a minimum strength to seed growth of structure.

Ever since 1967, we have had a theory that approached this problem.[1] The theory said that the background radiation must display a similar granularity to the lumpy universe, because gravity couples that early radiation to the matter. We do indeed observe this inhomogeneity in the universe, but at that time had no theoretical explanation. For one thing, structure did not have to be there, at least on the scales of the universe larger than clusters of galaxies. The presence of large-scale structure, however, implied that there must be an accompanying trace of large, angular scale fluctuations in the cosmic microwave background. One estimate in the 1970s came in at 10 percent, but this was soon ruled out.[2] At the time, we still did not have enough samples of the large-scale galaxy distributions for cosmologists to make a clear case.

A good number of physicists were studying the coupling and growth of primordial density irregularities and predicting strengths of the early fluctuations. In 1967, I was a graduate student at Harvard, and I too had argued that galaxies formed from infinitesimal seed inhomogeneities. I used this argument to predict temperature fluctuations of strength 0.03 percent on scales of up to a few arcminutes. Even this estimate turned out to be wildly overoptimistic. This was hardly my fault, however! The primary reason for eventually lowering expectations was the presence of a dominant density of dark matter, yet to be discovered. A breakthrough did eventually come in 1984 when we performed the first successful calculation in which the associated temperature fluctuations were ten times smaller than the earlier conjectures had led us to believe.

Whatever the strength of the fluctuations, we know that their

1. The growth of fluctuations in the expanding universe is calculated from the theory laid down in the pioneering work of the Russian physicist Evgeny Lifschitz in 1946.

2. These pioneering estimates were given in 1967 by Rainer Sachs and Art Wolfe at the University of Texas, Austin.

density needed to survive the effects of the radiation field before they entered into a growth phase, when structure begins to develop. The fluctuations need to survive because radiation tends to expand and dilute any baryon overdensity. Only the largest baryon overdensities can overcome this radiation drag. Logically, then, there must be a minimum scale of surviving density fluctuations due to the coupling with the radiation field. This further implies a corresponding minimum angular scale above which the temperature fluctuations could survive and be detectable. By this path of reasoning, cosmologists had a way to find out the level of the initial fluctuations. The inference of small angular scale temperature fluctuations provided a crucial missing link in the connection between the initial conditions and the formation of the galaxies. Now we needed to observe it to compare it with our calculations. If these fluctuations were not seen, the big bang theory was in trouble.

How was one to test such a theory, in the era before the advent of very large terrestrial telescopes and telescopes in space? Several generations of cosmic microwave background radiation experiments took place. For a prolonged period, the improved experimental limits were above the predictions of the progressively refined theory. Each time there was a major experimental improvement,[3] the theoretical hurdle was raised with the advent of more precise calculations.

The final theoretical refinements came in 1984 with the realization that the presence of dark matter made an important change to our understanding of the required strength of density fluctuations. The dark matter was assumed to be cold, for otherwise it would have the effect of suppressing fluctuation growth. Cold dark matter boosted fluctuation growth, and therefore, smaller fluctuations would have been required initially. This was the prediction made in 1984, and it was based on the reality of dark matter, which cosmol-

3. This happened with the pioneering attempts of Bruce Partridge in the 1970s, then of Juan Uson and David Wilkinson in 1981.

ogists had just begun to understand as being pervasive in galaxies and galaxy clusters.[4]

As noted earlier, dark matter must consist of weakly interacting particles. This is essential in order to understand why the dark matter dominates only in the outer parts of galaxies, and never in stars. The dark matter would have decoupled from the radiation field at a very early epoch. Dark matter today acts like a very cold gas of particles that rarely interact with each other or indeed with ordinary matter. Typically, the plasma of baryons, which is like an electrically charged (ionized) gas, is tightly coupled to the radiation field. Any growth of fluctuations is resisted by friction with the radiation, like swimming through treacle. However, once the universe was matter-dominated, the weakly interacting cold dark matter particles could move freely through the radiation under the attractive force of gravity, which allowed fluctuations to grow.

With further analysis, our predictions regarding the fluctuations came in at an order of magnitude or so lower than the early estimates. The goal now was to search for temperature differences of about 1/100th of a percent, over an angular scale of 30 arcminutes, roughly the angular size of the full moon.[5] The fluctuations are substantially lower on smaller angular scales where the damping plays a role. It was not long before the geometry of the universe and the cosmological constant were probed via these predictions.

But there were no data, just tantalizing upper limits. All of this activity resembled the building of theoretical castles in the air until a major breakthrough occurred in 1992, when the Cosmic Background Explorer (COBE) satellite verified that temperature fluctuations amounting to thirty parts in a million were present over angles in excess of 7 degrees.[6] Almost another decade passed before

4. The new predictions of temperature fluctuations arising from structure formation were made in 1984 by Nicola Vittorio and Joe Silk in Berkeley, and by Dick Bond and George Efstathiou in Cambridge.

5. This corresponds to what is called the first acoustic peak.

6. The COBE instrument that measured temperature fluctuations was the Differential

the fine-scale anisotropy predictions on subdegree scales were confirmed. Ground-based experiment and balloon-borne experiments provided strong confirmation of the elusive signal.[7]

Flatness of Space

Today we speak of the space in the universe as flat, or Euclidean, in the way we can measure lines or angles. However, this flatness was not always evident during these decades of debate about the initial conditions of the universe. There was, in fact, a strong prediction that the space of the universe was curved, and thus, that curvature could be measured in the sky.

If space is curved, meaning non-Euclidean, then light travels on curved paths, which we call geodesics. A universe with a curvature like that on the surface of a balloon acts like a convex lens. In this kind of universe, the angular size of the fluctuations is slightly shrunk. In the converse case—that of a curved universe shaped like a saddle or hyperboloid—the curvature acts like a concave lens. The angular scales are increased. Hence, we can calculate that the natural angular scale is the peak in the fluctuation strength that occurs for fluctuations that have undergone the maximum amount of growth in the early universe. When measured, this peak is about half a degree. Temperature fluctuations are strongest on this angular scale.

However, reality is never quite so simple. In the cosmic microwave background alone, connections exist between curvature and other uncertain quantities such as the age of the universe. In particular, one has to know the age precisely in order to infer the curvature of space, because in an older universe the surface of the last

---

Microwave Radiometer, built by a team led by George Smoot, who received the Nobel Prize in 2006 for this work.

7. The acronyms by which these pioneering experiments are known are TOCO (in Chile) led from Princeton, and BOOMERANG (CalTech) and MAXIMA (Berkeley) on balloons.

scattering of radiation is slightly farther away and so the scale of fluctuations shrinks. This mimics the effect of curvature. We can only sort out this degeneracy if we have an independent measure of the age of the universe. Fortunately we do, via Hubble's constant.

One of the most important results from the Hubble Space Telescope was an accurate measurement of the rate of expansion of the universe. This specifies Hubble's constant H, and the age of the universe is approximately equal to the inverse of Hubble's constant. The inferred age is 13.7 billion years. The Hubble constant as measured by one of the Hubble Space Telescope's key projects is 72 km/sec/Mpc with an uncertainty of less than 10 percent.

One can also express the Hubble constant as the inverse of 14 billion years. In other words, if the Hubble constant really is constant, the expansion would have started from a point of infinite density some 14 billion years ago. Of course, Hubble's "constant" is not really a constant as we go back in time, so other effects must have intervened.

Nevertheless, we know the age of the universe. So we can go back to the peak in the cosmic microwave background fluctuations at half a degree. We can now make the highly significant inference that the universe is flat. The theory of curved space, once seriously entertained, is now put on the shelf. The spatial curvature of our universe is close to zero. Space is Euclidean. Space is flat. Indeed, this is one of the most significant findings of modern cosmology.

CHAPTER 4
How Stars Form

BEFORE WE HAD even imagined the existence of other galaxies, we had observed a sky full of stars. The way stars form was first described for us by the English scientist Isaac Newton, founder of the theory of gravity in the seventeenth century. We have come a long way since then.

Each of us can imagine this process ourselves. First, picture a great cloud of gas. It is essentially all hydrogen, but it also has trace amounts of metals. These impurities provide a means for the cloud to lose energy by radiation. The cloud cools down, and eventually gravitational forces dominate over forces due to differences in gas pressure. Fragments condense out of the cloud, held together by their own gravity. Newton, using his theory of gravity, realized that this fragmentation into planet- and starlike bodies was inevitable. He explained the process in a December 10, 1692, letter to Richard Bentley, the English theologian:

> It seems to me that if the matter of our sun and planets and all the matter of the universe were evenly scattered throughout all the heavens, and every particle had an innate gravity toward all the rest, and the whole space throughout which this matter was scattered was but finite, the matter on the outside of this space would, by its gravity, tend toward all the matter on the inside and, by consequence, fall down into the middle of the whole space and there compose one great spherical mass. But

> if the matter was evenly disposed throughout an infinite space, it could never convene into one mass; but some of it would convene into one mass and some into another, so as to make an infinite number of great masses, scattered at great distances from one to another throughout all that infinite space. And thus might the sun and fixed stars be formed, supposing the matter were of a lucid nature. . . . But how the matter should divide itself into two sorts, and that part of it which is to compose a shining body should fall down into one mass and make a sun and the rest which is to compose an opaque body should coalesce, not into one great body, like the shining matter, but into many little ones; or if the sun at rest were an opaque body like the planets or the planets lucid bodies like the sun, how he alone should be changed into a shining body whilst all they continue opaque, or all they be changed into opaque ones whilst he remains unchanged, I do not think explicable by mere natural causes, but am forced to ascribe it to the counsel and contrivance of a voluntary Agent.

In fact, Newton ended up attributing star formation to what he believed was the power of God to act unilaterally in the universe. It took another English astronomer in the twentieth century, James Jeans, to describe the process of fragmentation from a more rational, and mathematical, perspective. Jeans wrote in 1902, "From the intrinsic evidence of his creation, the Great Architect of the Universe now begins to appear as a pure mathematician." He went on to lay down Newton's ideas in more precise physical terms:

> We have found that as Newton first conjectured, a chaotic mass of gas of approximately uniform density and of very great extent would be dynamically unstable: nuclei would tend to form in it, around which the whole matter

> would eventually condense. All celestial bodies originate by a process of fragmentation of nebulae out of chaos, of stars out of nebulae, of planets out of stars and satellites out of planets.

It finally was Sir Arthur Eddington who realized what it took to distinguish a star from a planet. If the earth was always cloudy, and we could never see the sun or night sky, would we ever have predicted the existence of stars? Eddington first posed this question, and he came to a dramatic conclusion: stars must exist. As he wrote in 1921:

> We can imagine a physicist on a cloud-bound planet who has never heard tell of the stars calculating the ratio of radiation to gas pressure for a series of globes of various sizes . . . so that his nth globe contains 10^n grams. Regarded as a tussle between matter and aether (gas pressure and radiation pressure) the contest is overwhelmingly one-sided except between numbers 33–35 where we may expect something interesting to happen. What "happens" is the stars. We draw aside the veil of cloud beneath which our physicist has been working and let him look up at the sky. There he will find a thousand million globes of gas nearly all of mass between his 33rd and 35th globes, that is to say between 1/2 and 50 times the sun's mass.

The logic behind Eddington's reasoning must have proceeded along the following lines. Fragments of all masses formed from the unstable parent cloud by the destabilizing action of gravity. But it took special conditions for the fragments to shine. They had to be not just opaque, as is the case for a rock or planet, but they had to be hot enough to have radiation play a role in pressurizing them against gravity. The radiation inevitably leaks out, and a star is born.

There was one more mystery to be resolved, however, before the distinction between star and planet could really be understood. What was the ultimate energy source that caused stars to shine? Eddington realized that the sun could not possibly be older than the earth unless there was a source of nongravitational energy. German American physicist Hans Bethe eventually discovered the energy source at the core of the sun: thermonuclear burning of hydrogen to helium.

The First Stars

We can only imagine the moment when the very first stars formed, bursting into flame and lighting the dark universe. But why did they grow to different sizes, and did the process change as the universe expanded? Tracking the process is easy enough. The main principle to be recognized is that stars not only formed as fragments that broke out of the cooling gas cloud, but they also grew by accreting, or drawing in, more gas from the surrounding environment. One way of following this process in the early and later universe is by looking at the role of the speed of sound. Astronomers of old once spoke of the "music of the spheres," and today that has a practical application. The speed of sound, which worked differently at the beginning of the universe than it did later, controls the rate at which stars accrete more material, and thus grow.

At the beginning of the universe, there was very little that could cool the heat of the gas produced by the big bang. There were few heavy elements, and the only coolant was the formation of molecular hydrogen, which still existed in only trace amounts. The lowest energy levels in such pristine gas are due to molecular rotation, and correspond to a very hot temperature of 512 degrees Kelvin. This is to be compared with the 10 degrees Kelvin attainable in nearby interstellar clouds through the lowest levels of carbon monoxide molecules due to molecular vibrations. The energy levels are excited by collisions between atoms and molecules, resulting in loss of thermal energy by radiation, but the cloud temperature in the first clouds, with just hydrogen molecules available as effective coolants, only went down to around 1,000 degrees Kelvin. At this heat, the sound speed was ten times higher than in today's gas clouds in the universe. The speed of sound determines how rapidly a developing star can accrete, or gather up, more material to grow. The accretion rate is proportional to the cube of the sound speed.

Back then, because of the heat, the accretion rate was a thousand times higher than we see in stars forming today. The first generations of stars, in other words, were massive stars, ranging up to a few hundred solar masses, though at some point, when the injection of energy stopped the accretion, they could not grow any larger. In general, however, these massive stars grew so rapidly that they exploded as supernovae.

Those explosions were a key step in further star formation. At the cores of the massive stars, heavy elements had been forged by nuclear reactions, and now these have been ejected into the interstellar medium. They mixed with the pervasive hydrogen, and provided a rich gas of elements that other stars could gather up in their growth process. As the explosion and enrichment process progressed, the gas in the universe was enriched to about a thousandth of the solar level. Hence, the cooling of molecular clouds became more effective. Now there were a sufficient numbers of heavier atoms and molecules with low-lying energy states to change the mode of star formation.

This begins the story of stars that, like our own sun, formed in a cooler universe. They are smaller, low-mass stars with longer lives, and thus they last long enough to form galaxies. The slow growth of these later stars is all connected to a slower speed of sound, now that temperatures are far lower than early in the universe.

In our present epoch of star formation, the gas clouds are cooler and filled with heavy molecules such as carbon monoxide, which has very low-lying energy levels. The molecules can be as cold as a few degrees above absolute zero, so the clouds can become cooler as their molecules release energy, in the form of photons, as they collide with other atoms and molecules. The gas can cool to a temperature of only 10 degrees Kelvin. The speed of sound, similarly, can be as slow as only a few tenths of a kilometer per second.

Regardless of when they formed, stars are rarely alone. They congregate in clusters. The fragmentation of the parent cloud results in many stars forming simultaneously and in a range of masses. Most stars that formed in the history of the universe have been dim counterparts of the sun, having only 10 or 20 percent of our sun's mass. These stars are dim, but they are very long-lived. They shine for tens of billions of years. As mentioned, the massive stars have shorter lifetimes (counted in merely millions of years), but they are highly luminous and can amount to a hundred solar masses.

The star clusters themselves take on a variety of forms. The most massive are the globular star clusters. These include some of the most beautiful objects in the universe. They are spheres containing millions of stars. In the center, the crowding is so extreme that one cannot resolve individual stars and sees only a diffuse blob of light. In a globular star cluster, the stars are all in orbit about each other. The system of orbiting stars is perfectly stable. The globular clusters are the oldest objects in our galaxy. Many formed long before the solar system and even before the Milky Way disk formed, more than 10 billion years ago. Globular clusters populate the halo of our galaxy.

While globular clusters are generally old, and their massive stars all died long ago, we do find some younger examples. These are seen in the Magellanic Clouds, our nearest galactic neighbors. Young globular clusters are also found in rare colliding galaxies. Their age coincides with the time elapsed since the collision, suggesting that they may have formed during violent encounters of huge gas clouds. Indeed, most of the stars in the halo may have formed in long-since-disrupted globular clusters.

Today, the situation is quite different. Most star formation occurs in galactic star clusters, which contain young, massive stars. These clusters amount in mass only to a few thousands of solar masses, and are found in the flattened disk of the Milky Way.

Our own solar system is a product of the heating, cooling, burning, and exploding of millions of stars over vast stretches of time. We—humans, the oceans, our planet, and indeed all planets—are made from the ashes of the stars. The solar system condensed out of an interstellar cloud about 4.6 billion years ago. About 2 percent of the mass of the cloud was in elements heavier than helium. These were synthesized in long-dead stars. We can reconstruct the changes in the abundances of the chemical elements once we acquire two bits of information. The key step lies in determining the yields of heavy elements from the different nucleosynthetic sources. The second requires data on abundance patterns in stars of progressively larger ages. In this way, we can study the evolution of the chemistry of the universe.

Imagine a master chef at work. She concocts a recipe from fresh ingredients. She adds a dash of spices and prepares the food. Some ingredients cook slowly, others instantly. All is mixed and the meal is ready. Building a star has certain parallels. The master astronomer's task is complicated by the fact that her ingredients are in remote parts of space. On her computer, she mixes the ingredients. These are the debris from dying stars.

A star shines because it has an interior source of energy. Most stars consume hydrogen, forming helium. The transformation releases about seven-tenths of a percent of the initial mass of hydrogen into energy. This keeps the center of the star hot, supporting it against gravity. The radiation diffuses out to the surface. The temperature of the surface layer of the sun is 6,000 degrees Kelvin. Consequently our sun is yellow. More massive stars burn more vigorously and are blue. The least massive stars are red.

When the core hydrogen is exhausted the core contracts. It heats up, and helium is ignited. The core is so hot and luminous that the outer envelope swells to form a red giant. As the helium burns into carbon, the outer layers are ejected, resulting in a planetary nebula.

The sun is destined to become a red giant in about 5 billion years, when the earth will be incinerated by the outflowing hot gases. The relic left in the center is a white dwarf, consisting of helium and carbon. It will have a mass of about 60 percent that of the sun, but will be so dense that it is smaller than the earth. The outflowing gas from the planetary nebula is the major source of carbon for the surrounding interstellar medium and eventually for planets.

A star more than ten times the mass of the sun evolves more rapidly and more strongly. It becomes a red supergiant. These giant stars are a hundred times more luminous and ten times larger than a red giant. The core continues to heat up, burning the carbon into oxygen, the oxygen into silicon, and finally the silicon into iron. Successive layers of the star contain each of these elements, rather like the layers in an onion. Once the core becomes iron, that is the end of the road for extracting stellar energy. No more nuclear energy is available via nuclear fusion. Iron is the most stable of the elements, the slag heap of the universe.

The final collapse of the iron core results in a star that is so highly compressed that the mass of a sun is contained within a sphere of a radius of only a few kilometers. This is a neutron star, in which the density is so high that elements lose their identity. A neutron star is a compact mass of neutrons. Enormous energy release is associated with the final compactification of the star. Conversion of the iron into neutrons produces a prolific outpouring of neutrinos, whose escape from the core is the reason the core loses energy and collapses. Most of the neutrinos are trapped within the outer parts of the star. The net effect is a vast explosion, or supernova. The supernova explosion results in the ejection of considerable amounts of oxygen, magnesium, and silicon into the interstellar medium. This is the fate of a massive star.

Another type of supernova explosion involves white dwarfs. A white dwarf has a maximum mass. If it were more massive, it would collapse into a neutron star. This critical mass, discovered by Indian American astronomer Subhramanian Chandrasekhar, is 1.7 solar masses. Now, many stars are in binary systems. It is not uncommon to have a low-mass binary system, each star being a little more massive than the sun. After some billions of years, we would have a white dwarf in a binary system. If the binary system was reasonably close to begin with—and many are—then, with time, the system contracts. The orbiting white dwarf loses orbital energy by emitting gravitational waves. It slowly sinks in toward its companion. The companion star in turn evolves and becomes a red giant star. The white dwarf may be close enough for it to capture much of the outflowing mass from its companion star, enough to push it over the brink, the Chandrasekhar limit. The white dwarf implodes, accompanied by a giant explosive release of energy. We have a delayed supernova. The delay is due to the merging timescale, which can take many billions of years. In this case, the debris from the explosion does not contain significant amounts of oxygen or silicon. But it is highly enriched in carbon and especially in iron, because the implosion is highly disruptive and does not leave a neutron star behind.

A very massive star, of more than thirty solar masses, terminates its life in a still more spectacular explosion. A black hole is left behind. The amount of energy in the explosion can be tens of times that in a supernova explosion. We call these objects hypernovae. Their explosions are believed to be responsible for the most luminous objects in the universe, the gamma ray burst sources. These objects provide a unique means of seeing massive stars in the very distant universe. Gamma ray bursts last a few seconds, but during this brief moment, the entire energy of a supernova is released. They are, briefly, the brightest objects in the universe.

Indeed, they are so bright, in the optical as well as in the gamma ray spectral region, that at least one was bright enough to be a naked-eye event despite being at a redshift of 2 (when the universe was twenty-seven times denser than it is today). Such violent explosions are believed to play an important role in the production of rare heavy atomic species whose origin requires more extreme and violent conditions than occur in supernovae.

All of these events, from red giant outflows to exploding stars, produce enriched debris that accumulates in the interstellar medium over cosmic time. All add up over billions of years to give the chemical composition of the nebula from which the solar system condensed. By studying the composition of stars of different ages, we can try to reconstruct the chemical history of the universe. We can best do this by studying the nearby stars.

Chemistry of the Solar Neighborhood

Stars are orbiting in our galaxy like cars on a circular race track. All have approximately the same speed. Some are slightly faster, some slightly slower. We call this variation the dispersion in stellar velocities. As time goes on, the stellar velocity dispersion inevitably increases. Massive clouds of gas orbit the galaxy. Stars pass by and receive a slight gravitational pull toward the cloud. These gravitational deflections add up cumulatively over time to heat the gas of stars. Astronomers can use the velocity dispersion as an indicator of age.

But velocity dispersion is not the only age indicator. There is also the iron abundance in the stars. Iron is generated by exploding stars and dispersed into the interstellar gas. The gas systematically becomes more enriched with time. In this way, iron also measures age because its abundance in the clouds out of which stars are forming increases with time. Stars with more iron formed later than stars with less iron. The velocity dispersion and iron abundance are correlated with each other. As one of these increases in a group of stars, so does the other—not too surprising as age is the fundamental parameter. But it does enable us to develop a coherent model of our Milky Way galaxy, especially in the neighborhood of the sun.

One surprise has emerged from this picture of systematic enrichment with time. There should be large numbers of metal-poor stars that are more primitive counterparts of the sun. Where are the stellar relics from bygone ages? Many such stars should exist, yet rather few are seen. Astronomers conclude that the galaxy has grown in mass. Perhaps half of the stars were formed after the interstellar gas already was enriched. The natural explanation is that the galaxy accreted substantial amounts of gas from its surroundings. This gas mixed with the interstellar gas and became enriched. In this way, the number of early-forming stars from unmerited gas was reduced.

It is not easy to deduce the chemical history when many of the stars have long since dispersed from the region where they were born. Star clusters provide a laboratory for decoding the history of the stars. All of the stars in a globular star cluster are old and were born together from the same parent cloud of gas. The chemical abundances of stars in a cluster provide a snapshot of conditions in the past, before even the sun had formed. Our picture of systematic enrichment is confirmed. We are literally formed from the ashes of stars.

The spectrum of a star reveals its chemical composition. Dealing with an entire cluster of stars is possible by taking spectra of individual stars, but we need to go much further: to study distant galaxies that are so remote that we cannot resolve individual stars. A recipe for decoding the chemistry of an entire galaxy consists of adding together all types of stars to produce a synthesized population of stars. The metallicity is adjusted until the synthetic spectrum matches that observed. Now we have reconstructed the stellar population, and its chemical abundances can be inferred. Thanks to stars, the secrets of the most distant galaxies in the universe have been unlocked.

CHAPTER 5
The Darkest Matters

EVEN THOUGH we do not know what dark matter is, we know it is everywhere. It fills galaxies, clusters of galaxies, and the universe itself. The evidence of its presence is convincing, albeit circumstantial. The culprit particle has not yet been identified, but physicists are desperately on its trail. At the least, we are confident that dark matter particles are weakly interacting compared to the proton, because dark matter mostly congregates on the outskirts of luminous galaxies and beyond. As seen earlier, dark matter helps us explain and calculate the early fluctuations in the universe. Now we want to use our hypothesis about dark matter to simulate the structure of the universe, and take our search to the great clusters of galaxies and the dark halos that surround galaxies.

DARK MATTER AND GALAXIES

Galaxy clusters provided the first hint of dark matter. Dark matter is required to explain how galaxies cluster and stay together. Although most galaxies seem sprinkled randomly through the universe, about 10 percent of them are in agglomerations—galaxy clusters. These contain hundreds or even thousands of galaxies. The question is how these clusters stay together, given their heat and energy, and given the speed at which they are moving as part of the expansion of the universe.

Galaxy clusters are our largest laboratories for studying the different components of matter. They allow us to take a census of matter, just as the census department of a nation counts its human population. Galaxies should be representative of the universe. Here is where the issue of dark matter becomes most apparent.

Clusters contain considerable amounts of ordinary matter that are not in the form of stars and visible in our usual telescopes. It took a new type of telescope to discover this. An X-ray telescope only observes from space, and this new technology was only developed in the last four decades of the twentieth century. Several times more mass is present in diffuse hot gas than is seen in stars. The gas is detected because it emits prolifically in X-rays. The gas temperature amounts to tens of millions of degrees Kelvin.

We also measure the hot gas by an even newer technique, exploited in the first decade of the twenty-first century. The gas casts a shadow on the cosmic microwave background at radio frequencies. In effect, the radio photons passing through the cluster are heated by the hot gas and are found in emission at higher microwave frequencies.

All of this gas, however, does not suffice to hold the cluster together. The galaxies stay together because of the mutual attraction of gravity. The gravity in luminous matter, both gas and stars, is not enough by far to maintain a cluster of galaxies. Galaxy velocities are measured from spectroscopic redshifts to amount to thousands of kilometers per second. The galaxies would fly apart within a billion years were not some unseen force maintaining them in orbit. That, we believe, is the gravitational role of dark matter.

If the binding together of a galaxy cluster is our first hint that dark matter exists, the rotation curves of individual galaxies are our most robust evidence for dark matter. A rotation curve is a measurement of the decrease in the rotation velocity of the galaxy as we move away from its center. Of course, at the center, there is no rotation. But stars orbit the center to set a rotational velocity that increases as we move out, but which eventually decreases as the gravity force weakens. In practice, we find that the luminous components of a galaxy, which are stars, do not account for all the mass necessary to produce the circular velocities measured in the outer parts of most disk galaxies. This is a large shortfall of mass. (We infer it to be about a factor of ten out to several disk scale lengths: one disk scale length is the distance out to which its brightness has dropped by 40 percent.)

A similar shortfall of mass occurs for elliptical galaxies. This mass distribution of ellipticals is traced by several techniques. One of the most direct ways to measure mass goes back to Einstein, who pointed out that gravity bends light rays. The images of background radio galaxies are actually distorted in shape by traversing the gravity field of the intervening galaxy, which acts in effect like a gigantic gravitational lens. (Other methods include mapping of X-ray emitting halo gas and determining gas pressure gradients, and the variation across the galaxy of stellar velocity dispersion.) All of these techniques demonstrate the dominant presence of dark matter on large scales, amounting to a hundred kiloparsecs or more in a given galaxy.

On larger scales, as we move away from the galaxy, the dark matter fraction rises until it reaches a value that is the same for galaxies everywhere in the universe. This universal value is found to amount to about 23 percent of the critical density on the scale of galaxy clusters (which is around one or two megaparsecs). There are several independent relevant measurements to verify this finding. (These include mass determinations of galaxy clusters from both the velocity dispersion of the member galaxies and the thermal pressure of the hot intergalactic gas.)

The total dark matter content of the great clusters of galaxies is now unambiguously measured by gravitational lensing of background galaxies. Lensing is the distortion of light rays produced by a gravity field. In this way one can test for the dark matter content of the intervening space. The image distortions trace the total matter content, and most of the matter is dark. In the inner region of the cluster, the lensing produces strongly distorted images. We refer to this as strong lensing. In the outer regions, the distortions are very weak, and it is necessary to average over hundreds of background galaxy images to bring out a signal, which we call weak lensing. With a combination of weak and strong lensing, one can use image distortions to trace the dark matter density profile out to several optical scale lengths, well beyond the optical confines of the galaxy distribution.

Where Are the Baryons?

Cosmologists speak of the universe as having a matter budget, the total amount of particles and energies that make up the mass of the universe. We can break this down into rough portions. The universe, for example, is 4 percent baryons (heavy particles), 24 percent dark matter, and 72 percent in a uniform component of dark energy. Detecting so much, or indeed any, dark energy was a great surprise, and we will say more about it later. Here we focus on the tangible components of the universe, its matter content, much of which is in galaxies and clusters. Furthermore, we can divide up the universe by what is luminous (visible) and what is dark, or not visible to human visual perception. For example, 10 percent of baryons are luminous, but 90 percent are dark. Dark means that matter is not luminous, but at least some of the dark baryons absorb light from distant galaxies, which makes them detectable.

While we have this general budget to go on, we are always trying to understand better where the individual portions reside. For example, where are the baryons? We know the dark matter content of the universe, and its value on large scales, by way of the presence of a dominant contribution of nonbaryonic dark matter. Independent evidence of a substantial nonbaryonic contribution to the matter budget comes from primordial nucleosynthesis of the light elements, which tells us that there are ten times more baryons in the universe than we see in stars. The successful interpretation of the baryon density comes from the comparison of the predicted abundances of helium, deuterium, and lithium with observational data. This comparison yields a baryon density that is only 4 percent of the density that defines the dividing line between an expanding and a collapsing universe. We call this reference density the critical density. At the critical density, the universe continues to expand forever. This is what we observe, and it is the dark matter and dark energy that add up to the critical density.

Most of these baryons are needed to account for the baryons observed in stars and in intergalactic gas. But the baryonic content falls far short of the total matter content of the universe. The non-baryonic content of the universe amounts to about six times the baryonic content. Precisely this ratio is measured in massive clusters of galaxies, where the intergalactic gas content itself is about three times the mass of the stellar content.

Lensing of individual galaxies allows one to compare the baryonic content, inferred by near infrared observations, with the dark matter content on galactic scales. Here one encounters an important clue that has implications for galaxy evolution and the intergalactic medium. Only about half of the primordial baryons are present in massive galaxies. Where are the remaining baryons?

The observed repository is the intergalactic medium. Vast intergalactic clouds of enriched gas are measured in absorption against distant quasars. Most likely, the gas has been ejected from galaxies after being enriched by early star formation. Energetic winds are indeed observed from massive galaxies in the early universe as well as from nearby star-bursting galaxies.

There is no great mystery about the baryon budget. Most baryons are dark, meaning they are not in stars, but are present as diffuse gas in the intergalactic medium. Inside clusters, the diffuse gas emits X-rays and is mapped with X-ray telescopes. About 10 percent of the baryons are unaccounted for. The intergalactic medium is probed, outside of galaxy clusters, by absorption line studies. Presumably the remaining baryons reside as a tenuous ionized gas in the depths of intergalactic space.

The Wider Universe of Dark Matter

We have just considered dark matter at the galaxy level, but now we move to much larger scales. The definitive measurements of the global dark matter density come from studying the dark matter distribution on scales even larger than those of galaxy clusters. Galaxy redshift surveys of up to a million galaxies probe the three-dimensional density distribution of the universe. One can venture out to a redshift of 0.2 for the 2DF and Sloan galaxy redshift surveys. By focusing on the old red and luminous galaxies, one can probe even more deeply. For example, one reaches out to a redshift of 0.5 in the case of the luminous red galaxy sample from the Sloan survey. Hubble's law is used to transform redshift into distance. The linear expansion law (with velocity proportional to distance) is a good approximation in the nearby universe.

However, the resulting map of the galaxy distribution seems highly distorted, because local velocities relative to the Hubble expansion are not easily distinguished from the uniform component of the expansion. All we measure is a redshift, which gives the total velocity of a galaxy away from us. One consequence is streaks in the map, always pointing toward the observer. These streaks are due to individual galaxy clusters and have been dubbed the "fingers of God." In addition to the recession velocity common to the entire cluster, an individual galaxy may have a local component of velocity projected toward or away from us. If toward us, this reduces the inferred distance in so-called redshift space. If away from us, this distance is slightly increased. The net effect is that of streaks pointing toward or away from us in the redshift space map.

These redshift space distortions allow a direct measurement of the dark matter density, since it is the dark matter that is producing the local velocities. This is not all the dark matter, since a uniform distribution of dark matter has no preferred direction for any distortion and leaves no trace in the redshift distortions. If uniform, dark matter exerts no influence on local velocity measurements. In effect, it pulls equally from all directions.

The measured dark matter density represents the global value of mass to light for the clumpy component. The inferred total density of dark matter is subject to correction by an additional factor because not all of the dark matter is inhomogeneously distributed. To arrive at the true value, we need to allow for the so-called bias factor, or the ratio of luminous to dark matter fluctuations.

On sufficiently large scales, there is no bias. Light traces matter. Hence a truly large-scale measurement, such as one involving the cosmic microwave background fluctuations, is needed to arrive at the true dark matter density. Luminous matter is not a good tracer of dark matter on smaller scales, such as those of galaxies, where the baryon density is enhanced because of cooling. Of course, the dark matter cannot lose energy by cooling, although one complication is that its density is enhanced by the contraction of the baryonic component, which has implications for galaxy formation.

When the measured three-dimensional galaxy distribution is smoothed over scales that average over the largest bound structures, which are the clusters of galaxies, the average matter distribution is found to be clumpy. The fluctuations in the galaxy distribution match those in the cosmic microwave background, scale for scale, once account is taken of the very different redshifts at which the two measurements are made. One can detect the "wiggles," or acoustic oscillations (that is, the analogues of sound waves), both in the background radiation and in the galaxy counts.

The amplitude of the wiggles measures the baryon density but the separation measures the particle horizon at last scattering. In effect, this scale specifies the curvature of the universe, which indeed turns out to be very close to zero.

The flatness of space has been confirmed both in the radiation and in the matter oscillations. For space to be flat, the universe must be at the critical density required for it to be just able to expand forever. This provides the strongest evidence for the universality of nonbaryonic dark matter. Luminous matter accounts for only half a percent of the critical density, and baryonic matter for 5 percent.

In fact, there is still a big puzzle, because the redshift space distortions only measure the total dark matter density. This is mostly nonbaryonic and amounts to 25 percent of the critical density. But there is much more. The remainder, contributing most of the critical density, is uniform as well as dark. It is dark because we don't see it, and homogeneous because it does not produce any additional redshift distortions. The redshift space distortions we measure are attributed to clustered dark matter. What is left is dark but uniform. It could be very hot dark matter, but if this were the case, structure could never have formed. Such matter has pressure, and pressure opposes gravitational attraction. If the gravity of the hot component were dominant, ordinary matter could not collapse on galaxy scales. The pressure of such a hot component would dominate over gravity at galaxy scales and even at cluster scales.

The other option is a dark energy field. This has negative pressure, rather like tension in a string. Negative pressure acts like antigravity, indeed causing acceleration of the universe on very large scales. Such an energy field is the preferred solution. The bulk of the critical density is contributed by dark energy. Dark energy has been measured by the acceleration of the most distant galaxies. We discuss the observational evidence for dark energy in chapter 7.

Matching Data to Theory

The quest to understand dark matter is teaching us a lot about the challenge of matching our theories to our data, and vice versa. Let's go back to the story of Arthur Eddington, who, by sheer logic, could conclude that there must be stars, even when the earth was hypothetically clouded over and he could not see them. Now, let us imagine a modern physicist who is today on a similarly cloud-bound planet. She is faced with the question of whether such a thing as a galaxy exists.

She uses her elementary knowledge of physics. Quite amazingly, she can indeed deduce that galaxies do exist, and they exist with masses typical of our Milky Way. This is the power of theory—at least in the beginning. She would have reasoned her deduction this way. First, a galaxy consists of a vast assemblage of stars that formed out of the fragmentation of a giant cloud of gas. Second, a necessary condition is essential for fragmentation into stars. This condition, she argues, must be that the cloud would cool effectively, presumably during the time that it is collapsing. Finally, she applies the proper cooling requirement as a limit on the mass of a galaxy and correctly obtains the size of her galaxy, say, 100,000 million solar masses.

But she now gets stuck. Theory making has its limits. Because smaller galaxy clouds cool more effectively, she would have next reasoned that there should be thousands of very small galaxies for every Milky Way. But this is wrong. More precisely, it is wrong if the dwarf galaxies contain stars and are luminous. There are some dwarfs but not nearly enough. Using theory, she might have also deduced that the inexorable pull of gravity meant that there should be some very massive galaxies, since the gas would eventually cool and be accreted onto a galaxy that has already undergone its initial collapse and star formation. This is wrong as well.

Finally, she makes one more prediction, again using theory. Galaxies form in a bottom-up fashion, so she now reasons, based on her figuring out that the simplest initial conditions for the fluctuations in the early universe amount to just one number. This is the measure of the strength of the fluctuations. If the strength is the same, the largest fluctuations begin to grow after the smaller ones, as it takes longer for light to traverse them and synchronize growth by gravitational instability. So she deduces that big galaxies form after small ones, and more slowly, since their collapse time is longer. This is wrong on both counts: the opposite is observed. The data tell the truth.

Let's take our imaginary physicist's score. She managed to succeed with one prediction out of five—namely, that a galaxy could exist. Still, this is a pretty poor record. It demonstrates the fallibility of our theories without corroborating data. In fact, the data always dictate the true story. Physicists are running hard to keep a step ahead of the data and make predictions for future experiments. And it only gets worse. When we ultimately consider the challenges of understanding dark energy, the conclusion is clear: We need a new physics, one that goes beyond our conventional approach to gravity.

The Dark Matter Solutions

A new physics may indeed reside in the dark matter. Until we detect it directly, we have to be concerned that it might be a manifestation of a more fundamental issue. Nevertheless, we can assume, for now, that dark matter will play a dominant role in our understanding of galaxy formation. Dark matter acts like a fabric that gravitationally controls the disposition of the baryons. Baryons are carried along for the ride as the dark matter becomes more concentrated. Eventually the dark matter condenses into discrete clouds.

The first objects to condense out of the expanding universe are small clumps and clouds of dark matter. Density fluctuations grow in strength and produce clouds. We map the temperature fluctuations in the cosmic microwave background radiation. The maps tell us that the fluctuations are stronger on smaller scales, which means that the growth process, via gravity which treats all fluctuations democratically, is bottom-up. The first clumps to form are equivalent in mass to one earth. These are not very useful for forming galaxies because such small clumps do not trap baryons when they form. The gas pressure of the baryons can only be overcome in clouds of mass a million solar masses or more. Once gravity dominates over gas pressure, the baryons are trapped and begin to lose energy by radiation.

When the universe was about a tenth of its present size, baryons were able to cool effectively for the first time. The resulting clouds of cold gas fragmented to form the first stars. Clouds of dark matter, gas, and stars merged together. Eventually, the massive clouds were formed that we identify with typical galaxies. The mysterious early fabric of dark matter made this possible.

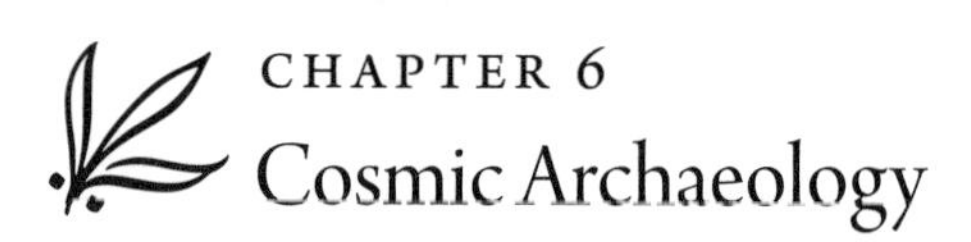

CHAPTER 6
Cosmic Archaeology

LIKE ARCHAEOLOGISTS, cosmologists find a range of puzzling objects as we sift through the billions of galaxies and stars. Many of these, naturally, continue to pose complex questions. How, for example, did the very unusual galaxies and stars form? We are also challenged to understand a process we call feedback, in which objects in the universe have interactions with other objects. This chapter looks at some of these fascinating puzzles. We'll begin with something called the cosmic web.

This web is actually a texture of filaments that penetrate the large-scale structure of the universe, and they are apparently related to dark matter. The cold dark matter paradigm has had considerable success in accounting for the large-scale structure of the universe. Its key ingredients are, for one, the collisionless nature of dark matter, and second, its primordial power spectrum of density fluctuations (which peaks at the 20 Mpc scale corresponding to the horizon at matter-radiation equality). This heady combination means that structure inevitably develops a filamentary structure on large scales. This is a consequence of caustics, a kind of wavy dynamic shape, first forming on the smallest scales that are nonlinear. This caustic effect is seen in swimming pools as they reflect light. In a region of space, the onset of a collapse sets a preferred motion. Motions in one direction encounter other flows. The effect is a causticlike behavior.

Caustics were first recognized as an ingredient of large-scale structure by Soviet cosmologist Yaakov B. Zel'dovich in 1970. He called them cosmic pancakes. They were sheetlike deformations of large-scale, one-dimensional converging flows. Zeld'ovich had an earlier life as a chemist and a nuclear physicist. He played a crucial role in the design of the thermonuclear bomb in the former Soviet Union. For this work, he was recognized as a Hero of the Soviet Union. His early career as a bomb designer made it difficult for him to get clearance to travel to the West. He only succeeded in temporarily leaving the Eastern bloc countries near the end of his life. Despite these obstacles, he was one of the world's greatest cosmologists in his lifetime.

Modern observations of the three-dimensional structure of the galaxy distribution have eloquently confirmed the filamentary pattern of the cosmic web. This term describes how massive galaxy clusters are situated at the intersections of filamentary structures. Thanks to accretion of diffuse matter and clusters, as well as galaxy halos, growth mostly occurs along the filaments. Two large galaxy surveys of the nearby universe have provided maps of the cosmic web. One is the Anglo-Australian two-degree field galaxy redshift survey of a quarter of a million galaxy spectra, which was superseded by the U.S.-led Sloan Digital Sky Survey of the spectra of a million galaxies.

Numerical simulations on computers complement the galaxy surveys. The largest of these simulations, the Millennium run performed at the Max Planck Institute for Astrophysics in Garching, Germany, included more than 100 million galaxies. The computer simulations followed 10 billion particles in the expanding universe and required several weeks of dedicated computing on one of the world's largest computers. The Millennium simulation was able to successfully reproduce the weblike structure of the galaxy distribution. Galaxies serve as markers of large-scale structure. The key assumptions underpinning this successful comparison of the virtual and real universe, apart from the fluctuation power spectrum, are that the dark matter is cold and interacts only weakly with the baryonic component.

Baryons account for 15 percent of the matter content of the universe and, unlike the weakly interacting cold dark matter, are able to dissipate energy, cool, and condense into galaxy-mass clouds. To form stars, the baryons must be able to cool within a collapse time, which we call the dynamical timescale. Cooling is a necessary condition for star formation. From this requirement, theory is able to account for the minimum and maximum masses of the luminous stellar components of the observed galaxies. Such remarkably simple arguments yield a minimum baryonic mass of 1 million solar masses and a maximum mass of 1 trillion solar masses for galaxies. These masses accord well with observations. There still are problems, however, with the predicted number densities, both of small and of massive baryonic objects.

The Dwarf Problem

Another cosmic archaeologist's question is about the very smallest galaxies—the dwarf problem. We should see lots of them, but in fact we don't seem to see enough. Theory tells us that the minimum mass of a dark mass clump within which baryons are able to cool is around 1 million solar masses. The gas can form stars, and so such clumps may be visible as dwarf galaxies. This hypothesis is appealing, as these clumps are plausible candidates for dwarf galaxies. Indeed, dwarfs are the building blocks for more massive galaxies. So far, so good.

However, there is a serious problem. Vast numbers of such dwarfs are predicted. One of the strongest objections to the cold dark matter hypothesis for structure formation is that the bottom-up scenario motivated by inflation and by the measured density fluctuations results in a huge number of small satellite galaxies. This predicted number greatly exceeds the number of satellite galaxies that we actually see.

Three possible solutions have been proposed for the shortfall in dwarf galaxies. One involves supernova explosions in the first dwarf galaxies. So much gas is expelled that little gas remains, and the number of stars is greatly reduced relative to the stellar fraction in a massive galaxy. Expulsion of gas is actually observed in starburst galaxies. The well-studied nearby examples of starbursts typically have energetic galactic winds and are associated with low mass galaxies. The outflow rate is observed to be comparable to the star formation rate. The outflows are driven by supernovae whose hot ejecta mix with interstellar gas and drive a wind. The same phenomenon in more massive galaxies results in a fountain. This spectacular event shoots out into the halo for kiloparsecs from a region in the disk of the galaxy. Here, gas is ejected from the disk, cools in the halo, and falls back into the disk after some hundreds of millions of years. However, for the small dwarfs predicted by dark matter theory, the galactic potentials are so shallow that after even the first supernova, one plausibly expects most of the baryons to be ejected.

Dynamical considerations restrict the numbers and masses of surviving dark matter halo clumps. There is a risk that the large number of surviving clumps may overheat the disk. Disks are observed to be thin, and their stellar components must be dynamically cold.

The second possibility for resolving the dwarf problem circumvents any issues of overheating the disk by appealing to tidal disruption of the dwarfs. Many clumps are completely disrupted by the effects of gravitational tides. These try to pull the clumps apart and result in their heating. Simulations suggest that surviving clumps have trajectories that are predominantly perpendicular to the disks. In this situation, overheating of the disks may not be an issue. The dark matter should consist of many tidal streams of debris, which would be merged into a quasi-homogeneous soup in the inner galaxy. Stellar tidal streams are indeed observed to form as relics of orbiting dwarfs in the halos of the Milky Way and of M31.

If these alternatives do not suffice, there is a third possibility. When the universe is reionized, the gas everywhere is heated to about 10,000 degrees Kelvin. This happened at a redshift of around 10, when the universe was a thousand times denser than today and before most of the dwarf galaxies had condensed. One consequence is that the smallest dwarfs are gas-free: the gas will be too hot to be retained by their weak gravity fields. This condition may be sufficient to break the hierarchy of gas-rich dwarf formation. A few will retain gas, perhaps in accord with the observed frequency.

We do not know which of these alternatives works best. Studying what happens to the leftover gas requires extremely high-resolution observations with large telescopes. Improved computer simulations are also needed. These are not easy because of the detailed physics that is required to study gas retention and star formation. Having several options would typically involve different types of computer simulation codes.

The Massive Galaxy Problem

There is a limit to the luminosity of a galaxy. This is not the case for a black hole. A galaxy is rather special in that its constituent stars need to form and shine. If galaxies were too compact, the stars would collide and disrupt. Galaxies cannot be too massive, or else the gas from which they formed would have been too hot to cool down and fragment into stars. Natural limits exist on the sizes and masses of galaxies.

One way we measure galaxy properties is by making large surveys and studying the statistical properties of the galaxies. The concept is similar if we survey people for a census. The galaxy luminosity function is observed to have an approximately exponential cutoff (at least for normal galaxies) at the bright end. This is difficult to explain. One might expect stars to form efficiently during the time it takes the accreting gas clouds to fall into the galaxy. This is certainly true, but infall and cooling during a dynamical time are only necessary requirements for stars to form, not guaranteeing that star formation happens.

Baryons continue to accrete onto massive galaxies, and there is no compelling reason for star formation to stop after a dynamical time. Indeed, theory suggests that the baryons should accumulate, cool, and continue to form stars, which means that massive galaxies are expected to still be forming stars and hence still growing in total stellar mass. Yet the most massive galaxies are ellipticals, galaxies that are almost completely devoid of ongoing star formation and whose space density is fit by an exponentially decreasing luminosity function. Growth ceased long ago, apart from the occasional merger. This does not conform with expectations from simple theory. We need to quench recent star formation in massive early-type galaxies. In dwarf galaxies, supernovae play a quenching role. But in massive galaxies, supernovae cannot do this. The gravity field is simply too strong. A more powerful and coherent source of energy is needed.

The dilemma of the massive galaxy problem has been resolved via an unexpected addition to the cosmologist's toolset. Supermassive black holes (SMBH) provide the new ingredient. These are lurking in the centers of most galaxies. They are detected by the presence of a highly concentrated dark component of matter. Ordinary stars cannot account for the local matter density at the center of a galaxy such as our Milky Way. The only explanation is the presence of an extremely massive black hole. In the case of our galaxy, the mass required is 4 million solar masses. We also see galaxies with much larger black holes. For example, in the elliptical galaxy Messier 87, the central black hole weighs in at 3 billion solar masses. Other black holes are known, most typically a few solar masses. These are the remnants of massive stars. Understanding how SMBH formed is an unsolved problem that we speculate about below.

Today the supermassive black holes are generally quite inert, apart from rare exceptions. However, these rare exceptions, which are black holes that still are feeding on the material in the universe, provide glimpses of what might have happened in the past. Black hole feeding is rare today. It is unusual to find active examples in the nearby universe. But this was not always the case. In the past, feeding was common. There was a frenzy of accretion and merging. Indeed, accretion is how galaxies accrued their mass. This same accretion would also feed black holes.

At one time, there was an intimate connection between supermassive black holes and galaxy formation. We know this because the more massive the spherical component at the center of the average galaxy, the more massive is the central black hole. An empirical correlation is observed between the masses of supermassive black holes, universally found to be at the centers of galactic spheroids, and the stellar masses of the spheroids. The correlation applies to spheroids similar to the bulges of the Milky Way and of the Andromeda galaxy, all the way up to the most massive galactic spheroids at the center of elliptical galaxies such as Messier 87. The ellipticals are pure spheroid in structure whereas disk galaxies are a hybrid of disk plus spheroid. The empirical correlation applies generally to all spheroids. Spheroids contain only old stars. Spheroids formed first, before disks. We infer that virtually the entire phase of supermassive black hole formation and growth must have coincided with spheroid assembly during the epoch of galaxy formation. Disk assembly occurred later by gas accretion.

Supermassive black holes are associated with an intensely active but short-lived phase at the centers of galaxies. The black holes themselves are invisible. However, they accrete gas from their surroundings. The gas heats up as it falls into the black hole. Because the gas has angular momentum, a central disk forms. The disk continuously accretes at its outer edge, and infall to the black hole occurs as the disk shrinks and inner bits break off and fall into the central hole. The matter becomes so hot that energy is injected beyond the disk, which usually happens by way of twin jets that stream out along the axis of the rotating disk. The surrounding cooler gas is energized. The net effect is to produce a luminous central object that is seen as a bright nuclear source. Active galactic cores are powered by energy outflows from supermassive black holes. These active galactic cores, or nuclei, are observed at as high a redshift as quasars are observed, and in some cases quasars are identified with the cores of host galaxies. The core of nearly every galaxy thus once underwent a quasar phase, which commonly happened at the epoch of galaxy formation. Galaxies and quasars must have formed simultaneously. The supermassive black holes are sometimes refueled by a galaxy merger, but these are rare in the recent universe. So are quasars. The most distant quasars are among the most luminous objects in the universe.

The hot gas glows, and the light spectrum is seen to contain spectral lines, which enable the strength and speed of the outflow to be measured. Vigorous outflows are observed from the core emission line regions. The outflows generated amount to a solar mass or more per year. The gas flows move at up to a tenth of the speed of light. The outflows interact with ambient interstellar gas clouds. The detailed interaction depends on the rate at which clouds are shredded and dispersed. The complex interactions need to be simulated numerically on the largest computers available, since it is necessary to resolve microscopic scales. The basic conclusion is that quasar outflows energize their surroundings. The transported momentum is found to have a substantial impact on the surrounding interstellar medium. Indeed it suffices to drive out the bulk of the diffuse interstellar medium from the galaxy.

We actually observe the phenomenon of gas expulsion. Researchers study absorption lines produced by the outflowing gas seen against the central galaxy. Global winds are mapped out. These emanate from ultraluminous star-forming galaxies that often have active cores. The outflow rate is a significant fraction of the star formation rate. The outflows can be as much as a thousand solar masses per year.

The outflows surround and overtake obstacles such as interstellar clouds. The clouds are crushed. Small clouds are evaporated and dispersed. Larger clouds are induced to collapse. The massive dense interstellar clouds that are engulfed by the outflows will be overpressured. They are induced to collapse and form stars. The resulting supernovae will add to the driving of the outflow. It is likely that the outflows are strongly enhanced by triggered star formation.

Less massive clouds are expected to be shredded by gas instabilities. These act much like the mixing of cream on stirring a cup of coffee. In contrast, the massive clouds survive and provide new stars, and especially some massive stars that rapidly explode as supernovae. Observers find that most luminous infrared galaxies indeed have both strong nuclear activity and star formation, which confirms that the two phenomena are closely connected.

The masses of the supermassive black holes are studied both in the distant as well as in the nearby universe. The host galaxy masses are also measured. The conclusions are discrepant. Most studies maintain that the supermassive black holes are correspondingly more massive than the stellar components in the early universe relative to the present-day observations. If the faraway black holes are indeed overly massive, one would infer that they formed first. If, on the other hand, one found in some cases that the black holes were undermassive relative to what one finds nearby, one might infer that the spheroids of stars formed before the black holes were in place. No doubt the truth is a compromise between the alternatives of advanced or retarded black hole growth. One cannot say whether supermassive black hole formation and growth precedes that of the spheroid or is a sequel. It all depends on whether active galactic core feeding and outflows drive star formation or vice versa. No doubt the truth will be elucidated by future observations, especially at far infrared and X-ray frequencies.

Star Formation in Galaxies

Despite extensive research on stars, models of their formation are based only on phenomenology—that is, a description of what we see. In terms of a deeply satisfying and fundamental theory, however, we really do not understand either star or galaxy formation. In our phenomenological approach, we study nearby cold gas clouds. These contain mostly molecular gas and dust. These are the sites of star formation. In the coldest, darkest clouds, one can find evidence of very young stars by searching in the far infrared spectral region. As stars age, they heat up their surroundings. The dust reradiates the starlight in the infrared spectral region. Dark clouds often contain embedded infrared sources. These are forming stars, still surrounded by the opaque embryonic gas and dust from which they condensed. Only after millions of years have elapsed does the curtain lift. Then the dust disperses, and we directly view the newly formed stars.

Star formation is the key ingredient in understanding how galaxies formed. Our knowledge of stars is based on local data, which provides the basis for phenomenological modeling of the initial mass function and the star formation rate and efficiency. However, whether these models apply to the extreme conditions of the early universe is far from certain. Brute force simulations cannot suffice, hence analytic descriptions are essential. What one does know is that two modes of star formation most likely are required to explain the two distinct types of stellar populations. One mode of formation is red, old, and gas-poor, and has the shape or morphology of a spheroid. A second mode is blue, young, and gas-rich, and populates the disks. Young, gas-rich, star-forming disks are currently forming stars with low efficiency. Inside the old, gas-poor massive spheroids, stars formed long ago with high efficiency. This is the first mode, and it helps explain the formation of galactic disks. The second mode is responsible for elliptical galaxy and spheroid formation.

Star Formation in Rotating Disks

Stars still are forming today, and it seems to be an inefficient process. The natural timescale for the accumulation of the gas, the raw material for star formation, is what we call a dynamical time, which is the time it takes a galactic particle to fall into the galaxy from far away. It is also the orbital timescale. In the case of our Milky Way, this time is about 100 million years. Our galaxy contains about 6 billion solar masses of cold gas. If this gas all formed stars in a dynamical timescale, we would be experiencing at least ten times the observed star formation rate. In fact, the situation is even worse because the gas is mostly in cold, dense clouds. This would predict that we'd see even more stars. These cold, dense clouds collapse on an even shorter timescale of at most 10 million years.

We infer that the fraction of the gas supply expended per global dynamical time is small. Star formation in typical clouds must be highly inefficient. Star formation is generally observed to be a prolonged process. In the Milky Way galaxy, star formation has been continuing for 10 billion years. The same is true for other disk galaxies with spiral arms. The longevity is due to the low efficiency but even more importantly, to the availability of a fresh supply of cold gas. The cause of the inefficient utilization of the gas supply is due to supernova energy. Supernovae, exploding stars, drive momentum into the interstellar medium. Star formation is not always a process that gradually declines with time as the gas supply is exhausted. Gas refueling can occur, especially in lower-density environments where there are copious supplies of cold intergalactic gas. Indeed, in some dwarf galaxies, there are vigorous outbursts of star formation. In these starbursts, one sees dramatic outflows that are most likely driven by supernovae.

We have a reasonable overview of how star formation proceeds in disk galaxies. The disks are cold and gas-rich. They are supported by rotation. The origin of the rotation in disks seems clear. The halos within which the gas accumulates have a small amount of rotation. This comes from random gravitational interactions with similar mass neighbors. Acquisition of a small amount of spin is inevitable. The spin is shared with the baryonic gas. The gas cools and condenses. It spins more rapidly as it contracts since angular momentum is expected to be conserved. Eventually, a disk is formed that is supported in the radial direction by rotation.

A rotating disk that is cold and supported by its own gravity is unstable and thus prone to break up. Disk instability leads to fragmentation into cold, dense molecular clouds. The clouds orbit the galaxy and sweep up gas. The mass buildup results in the clouds themselves becoming unstable. Cloud collapse is inevitable. The clouds fragment into stars. Some of the stars are massive and explode as supernovae. The rate of star formation is itself self-regulated by supernova explosions. To continue fragmenting and forming stars, disks must be kept gas-rich and cold, and hence unstable. Continuing cold gas infall and accretion are essential.

What is the origin of the gas? In fact, galaxies are surrounded by gas reservoirs. The greater problem is why many galaxies are not forming stars. These so-called red and dead galaxies are the elliptical galaxies. In the galactic halo, there is a reservoir of gas clouds. Cooling gas is also expelled to the halo by minor mergers of gas-rich dwarf companions. Gas infall is essential for at least two reasons: it is needed to maintain star formation, and it is also essential for understanding the distribution of the chemical elements in old stars.

Disk galaxies grew by gas infall. Early on, the nascent disks had no metal-providing stars from which to draw material, and so gas was their main building block. Disks grow by gas infall. We know this because, otherwise, we would see many metal-rich stars relative to their metal-poor counterparts. We also infer that the infalling gas was relatively pristine, that is, unenriched in heavy elements by stellar processing. Fed by gas, the disk galaxy continued to grow in mass until it reached its present size. Thus a few billion years ago when the galaxy was metal-poor, there were relatively fewer stars being formed.

Star formation in a galaxy and the galaxy's cold gas content are connected by an empirical law, named after its discoverers Martin Schmidt and Robert Kennicutt. The Schmidt-Kennicutt law provides a striking phenomenological representation of the connection between star formation rate and the presence of a cold gas reservoir. This law says that the rate of star formation in the disk is proportional to the gas mass divided by the orbital timescale. The star formation efficiency, or fraction of stars formed per dynamical time, is observed to be about 2 percent.

There are two ways of understanding this number. Remarkably, supernova momentum input provides an efficiency of a couple of percent. So also does internal self-regulation of giant molecular clouds by massive star formation. The orbital timescale depends inversely on the square root of the density. The higher the density, the shorter the timescale. Astronomers find that all star-forming galaxies satisfy this relation. Nuclear starbursts, as well as star-forming complexes and normal star-forming galaxies, fall on the same law. The dependence of star formation rate per unit disk area on gas surface density is to the 3/2 power. If it were only a linear dependence, the star formation rate would increase by 100 for a 100 times increase in density. But the dependence is stronger. The collapse of gas is more rapid at higher density and enhances the rate at which stars form out of a given supply of gas. Thus, if the gas density increases by 100, the star formation rate goes up by 1,000. Star formation becomes a run-away process that we observe in the form of starburst galaxies.

Many questions remain to be answered. Why is there such a universal relation between star formation rate and total gas content that applies regardless of disk type? Why does this same relationship encompass starbursts? And there is a second empirical relation for disk galaxies, whose connection to the first law needs to be understood. This law is named after astronomers Brent Tully and Richard Fisher. The Tully-Fisher law connects galaxy stellar mass with rotational velocity. The rotation rate of a disk galaxy is controlled by the dark matter mass of its halo. The law asserts that the stellar luminosity of a galaxy is proportional to the fourth power of its circular velocity. As the circular velocity is doubled, the luminosity increases by a factor of sixteen.

We also need to ask, how is the Tully-Fisher relation to be accurately understood? In particular, why is there an apparent conspiracy between the stellar disk and the self-gravity of the dark matter potential? The dark matter controls the disk rotational velocity. And why do gas-rich dwarf galaxies lie on the same Tully-Fisher relation as do massive disk galaxies? This is surprising, since chemical evolution models suggest that they have indeed lost substantial amounts of gas via outflows.

There are more disk enigmas. According to our theories, the contraction of the gas forces a contraction of the dark matter halo due to the increasing gravity field in the inner regions where baryons dominate. As the halo becomes more condensed, the rotational velocity of the disk changes, according to theory, by increasing the velocity where the light matter is most concentrated. We can observe this. However, we also observe the converse. Even where the light matter falls away in the outer part of the galaxy, the rotational velocity remains high. In this outer region, luminous matter and the dark matter somehow must conspire to maintain a constant rotational velocity, even at these distances from the radius of the galaxy. All disk galaxies show this constant rotation, but how the conspiracy works remains a mystery.

Our numerical computer simulations of gas contraction and disk formation have uncovered another surprise. Most of the angular momentum of the gas is lost to the dark matter. The dark matter cannot collapse because it does not lose energy. It stays in a halo that acquires a little spin. Meanwhile, the baryons, which only amount to 10 percent of the dark matter, end up forming too massive a slowly rotating spheroid and too small a disk.

On the topic of disk galaxies, something is still missing from our detailed understanding of how galaxies form. Astronomers are convinced that the difficulties lie in the computer modeling of galaxy formation. Implementation of feedback into the gas is poorly understood. Feedback is the effect of supernovae in heating the gas and slowing its contraction before it forms most of the stars. Feedback is a complex process that involves supersonic turbulence and magnetic field regulation over a wide range of gas densities. We simply have not yet modeled feedback with enough precision to be confident of the outcome.

Star Formation in Spheroids and Mergers

Spheroidal galaxies provide even more serious uncertainties regarding our modeling of the formation process. Our problem here is exacerbated because ellipticals formed far away and long ago. The evidence that can explain their formation is indirect. Again, there are simple relations between measured properties of the galaxies that contain hidden clues. One such relation is the fundamental plane of elliptical galaxies. This plane connects three measured quantities: the size, the spread in velocities of the stars, and the mass of all the stars in the galaxy. A plot of these three quantities in three dimensions shows that they lie on a two-dimensional plane, which means that they are correlated. Specify any two, and the third is inferred. How this came about is a mystery. What is certain is that elliptical galaxies as observed today are very slowly evolving systems. Any correlation in their properties must have been imprinted at birth.

Galaxies today are very slowly evolving systems. These scaling relations must have been imprinted long ago, when the galaxies formed. The relations provide empirical constraints on how the stellar mass was assembled. We would dearly like to compare our snapshots of the past, deduced from the scaling relations, with formation models. However, unlike the case for disk galaxies, we have no fundamental theory of spheroid instability and star formation. Rather, there are phenomenological hints that suggest the gas reservoir for forming stars is controlled by mergers. Mergers without gas are called dry mergers. A merger with gas is wet. A wet merger involves star formation and leads to complex changes in morphology. Modeling mergers tells us that dry (or dryish) major mergers certainly played a major role at a redshift of 1 or 2 in accounting for the morphological properties of massive ellipticals.

However, major mergers are conceded not to be the primary cause of spheroid formation. There are simply too few of them at very early epochs. Environment clearly played an important role. One has to also confront gas-rich minor mergers as the feeding source for the collapse of the rare objects that are massive elliptical galaxies. A minor merger means one in which the mass ratio is less than about 1:10. These must be one of the primary routes for instigating spheroid formation. Mergers of gas-rich substructures supply the gas that forms the stars. Most importantly, the violence of the merger compresses and torques the gas. The gas falls into the central region of the forming galaxy. Once enough mass in gas builds up, star formation commences in earnest.

The formation timescale can be probed by studying the chemical properties of the stars. Elements like magnesium and oxygen are produced by massive star supernovae. These stars only exist during the first 100 million years of a galaxy's lifetime, if there is no further substantial injection of gas. The magnesium in stars was synthesized early in the galaxy's lifetime. However, elements such as iron are also made in supernovae, but mostly in supernovae whose precursors are very long-lived, low-mass stars. Several billions of years are thus required to accumulate the observed iron in stars. The observed iron was synthesized late in the galaxy's evolution. The ratio of magnesium to iron in the old stars is an effective archaeological clock. It tells us how much time elapsed when these stars were being assembled. The observed ratio in the sun requires the sun to have formed about 4.6 billion years ago. An elevated ratio means that the stars formed over a much shorter timescale. In this way, we learn that massive elliptical galaxies must have formed within 100 million years. Disk galaxies, by contrast, formed over several billion years.

It so happens that the collapse time of a massive galaxy was around 100 million years. From this data, we infer that massive elliptical galaxies formed most of their stars more or less monolithically, during the initial collapse. Of course, the gas must have accumulated for a while before forming stars. The enhancement in the magnesium-to-iron ratio leaves us with an unexpected conclusion. We had previously assumed that the bigger the galaxy, the later it formed, which is certainly true for dark halos. A corollary is that the later an object condenses out of the expanding universe, the lower the mean density, and the longer its initial collapse time.

I have just summarized the theory, but in fact we see the opposite. The data tell us that the gas component formed stars much more rapidly, on a timescale on the order of the dynamical timescale of a massive galaxy. Not much star formation could have happened before, when the gas was being assembled from many smaller objects. What inhibited the gas from making stars prematurely, before the massive galaxy was in place? The coincidence between star formation time, deduced from the chemical evidence, and collapse time, deduced from size and mass, is sufficiently compelling that monolithic collapse is favored. But this poses the urgent question: why monolithic, if the gas is assembled by minor mergers and accretion? Some new ingredient is needed.

As a solution, I argue that the monster in the middle, the supermassive black hole in the core of the galaxy, makes a difference. Supermassive black holes are found at the centers of galactic spheroids, which are observed as the active nuclei of galaxies. They accrete gas from their surroundings and are often the sources of tremendous outflows. Supermassive black holes could provide the missing link needed to understand the mystery of elliptical galaxy formation. The ultimate determinant of whether a galaxy contains a massive spheroid may well be the role played by the central active nucleus of the galaxy.

The evidence for such a missing link stems from a discovery originally made by astronomer John Magorrian. He discovered a relation between the supermassive black hole mass and the mass of the spheroid. The relation only makes sense if both formed together at more or less the same time in the early universe. Detailed measurements of spheroid velocity dispersions and black hole masses have provided eloquent testimony to an intimate connection between active galactic nuclei and spheroid formation.

Inefficiency in Disk Galaxies

We have seen that star formation in disk galaxies is an inefficient process. Consider how inefficient it is: the characteristic timescale for star formation, for example, is several gigayears for a galaxy such as the Milky Way. This is an order of magnitude longer than the orbital time of a star, which is the same as the dynamical timescale of a star. This slowing-down of the evolutionary clock is a consequence of what we call feedback from ongoing star formation. The feedback acts on the gas that has yet to form stars.

Feedback operates as follows. Stars form in a range of masses. The more massive the star, the faster it evolves. Stars of more than about ten solar masses die spectacularly within a few million years. Massive stars expend energy prodigiously and soon exhaust their nuclear fuel supply. The inevitable result is that the core collapses, releasing so much energy so rapidly that the rest of the star explodes. Radioactive debris is ejected that decays into stable products such as iron, carbon, and oxygen. The hot debris is swept up into an expanding shell that eventually is halted by the ambient interstellar pressure. The deceleration of the dense shell by the diffuse interstellar gas is an unstable interaction. Fingers of dense, cold gas penetrate into the light warm gas. The shell breaks up and disperses. The enriched debris from the explosion mixes into the interstellar medium.

The interstellar clouds are enriched. Processing of gas into stars is slowed because the supernova debris also inputs momentum into the clouds. The cloud gas is accelerated. This is not the only source of feedback. The rotation of the galaxy also stirs up the cloud motions. The galaxy rotates at a constant velocity. When a cloud periodically wanders into an inner orbit, its velocity increases relative to neighboring clouds, because angular momentum is conserved. A stone on the end of a spinning string has a larger velocity when the string is shortened. These shearing motions also accelerate the clouds. The energy source is the overall rotational energy of the galaxy.

All of this is feedback. There is even feedback from forming stars. The interstellar gas is threaded by magnetic lines of force. Clouds spin and wind up the fields as they contract, and eventually the field lines snap. Magnetic energy is released into the gas. Assembly of gas is slowed. Star formation is delayed. The lifetime of a molecular cloud is increased, up to an order of magnitude longer than the free-fall time. The observed gas turbulent motions clearly must dominate the gas pressure; cloud internal motions are observed to be highly supersonic. The turbulent pressure counterbalances gravity. Since the turbulence mostly is driven by stellar feedback, cloud star formation rates are self-regulating.

Global self-regulation is controlled by the star formation timescale and by the gas supply. Star formation depletes the gas supply. Dynamical interactions between stars and molecular clouds heat up the young stellar disk. Maintenance of global disk instability requires a cold disk, and the gas supply to maintain this is provided via gas accretion, which comes about by continuous infall from the halo as well as by minor mergers. All of these dynamic interactions enhance the gas turbulence. The turbulence is reflected in the cloud internal velocity structures. Supersonic turbulence is believed to play an important role in accounting for the longevity of star formation in disks. However, the global driving of the turbulence, which normally would dissipate on a short timescale, is not fully understood. Numerical simulations of entire disks generally do not resolve individual star formation and stellar explosions. Simulations have difficulty in following the dissipation of the turbulence.

The problem we have been discussing regarding disk galaxies

provides an insight into how feedback in the universe works. Here are some simple analytic considerations. The low efficiency of disk star formation lends itself to a simple interpretation of the fraction of gas converted into stars per disk rotation time. We have already noted that this is observed globally to be about 2 percent. The fraction of initial supernova energy available for stirring up the interstellar gas, after allowing for radioactive losses, is the ratio of the observed interstellar gas cloud velocity dispersion to the velocity at which an expanding supernova remnant has the effect of a snowplow in pushing all before it. A remnant expands initially at twenty thousand kilometers per second. Only when it slows down by sweeping up ambient gas to around four hundred kilometers per second does it effectively communicate its momentum. The ratio of cloud motions to the velocity at which momentum is first effectively conserved happens to coincide with the observed inefficiency of gas conversion. This is not a surprise. It is expected if cloud turbulent velocities are indeed maintained by supernovae.

Supernova-driven bubbles of hot gas that sweep up cold gas into a dense shell provide us with a self-regulation model for disk star formation. The self-regulation is controlled by the interplay between the cold gas reservoir needed for star formation and the negative feedback associated with massive stars, which includes both their birth and death. One complication is that the disk is not a closed environment. Supernova explosions trigger venting of disk gas into the halo. The gas cannot escape from massive disks. This is true for disks similar in mass to the Milky Way galaxy. The venting results in galactic fountains. These galactic fountains eventually cool and provide part of the reservoir of cold gas that maintains disk instability.

As a disk galaxy, the Milky Way galaxy is one of the best-studied examples in terms of baryon content and star formation history. There is a precise prediction for the baryon content of the universe from the synthesis of the light elements in the first three minutes of the big bang. We measure the mass in interstellar gas and in stars today. It is clear that known baryon reservoirs only account for at most three-quarters of the initial baryon content of the universe. As we asked in an earlier chapter, where are the baryons today? One possibility is that the halo contains compact massive baryonic halo objects (Machos), for which there is some, at best, marginal evidence. Another possibility is that most of the missing baryons are in the form of diffuse, million-degree Kelvin gas.

The cold dark matter theory of structure formation asserts that there is continuing accretion of cold gas. The details of how the gas heats up on interacting with the halo is currently debated. However, the current gas content is expected to have doubled the total baryon content over the past 10 billion years, which means that a typical galaxy should contain on the order of 100 billion solar masses in diffuse gas. This is a vast overestimate compared to what actually is observed.

Observational constraints require that this gas must be ionized, otherwise we would already have seen it. The gas acquires high velocities when it falls into massive galaxies. It must shock and heat up to around 10 million degrees Kelvin. At this temperature, diffuse gas glows in X-rays. X-ray observations of nearby massive disks such as the Sombrero galaxy find only around a billion solar masses in hot gas in the bulge region. The only explanation seems to be that supernova heating has driven most of the gas into the outer galaxy.

This supernova heating produces winds, and indeed such superwinds are found around Milky Way–type galaxies. This is puzzling from the theoretical perspective because supernovae are considered to be incapable of driving a strong wind from the gravitational potential well of a massive galaxy. Also, at least one example is known of an extended X-ray halo around a normal massive spiral galaxy. In this case, gravitational accretion and shock heating provide the most plausible source for heating the gas.

The situation for the Milky Way galaxy is complicated, because the Local Group (which is the local aggregation of galaxies dominated by the Milky Way and the Andromeda galaxy) provides a potential buffer against gas infall. An explanation of the paucity of halo gas in the Milky Way may lie in a complex history for the gas. Ejection plays an essential role. One possibility is that substantial amounts of gas were ejected into the Local Group intergalactic medium. This could not have happened recently, given the strength of the Milky Way gravitational field.

Ejection, however, could have happened long ago, when our galaxy was being built up from mergers of many smaller systems. Ejection of gas must have happened to some extent during the assembly phase of the Milky Way. Winds were easily driven from the star-forming dwarf galaxies that merged into the Milky Way. Some of these building blocks still remain in the halo as low-surface-brightness fossil relics. To further complicate matters, we know that the merging history of our galaxy itself involved dwarfs that were still gas-rich. Star formation and enrichment occurred during the merging process. We infer this because the chemical abundances of the nearby dwarfs show similar trends with metallicity as do Milky Way halo stars, which means that a universal connection exists between abundances of elements such as magnesium with that of iron. There is one difference, however. The dwarf chemical abundances are generally offset to lower metallicities. Presumably, then, dwarf formation occurred earlier than disk and stellar halo formation.

Much of the enriched gas eventually formed the disk. On the order of 50 percent could plausibly have been ejected by supernova-driven winds from the dwarfs into the intergalactic medium. In nearby dwarf starbursts, we indeed find that the wind mass ejection rate is comparable to the star formation rate.

More generally, low-surface-brightness dwarf galaxies are ubiquitous. Supernovae are capable of ejecting much of the initial gas content. However, this gas has to be respelled by accretion. Otherwise these systems would not lie on the observed trend that relates total stellar and gas mass to the rotation velocity of the system (the Tully-Fisher relation discussed previously).

Surprisingly, the low-surface-brightness dwarfs are on the same relation as their more massive disk counterparts. Baryon-rich, star-poor dwarfs are indistinguishable, in terms of the Tully-Fisher correlation, from their gas-poor but star-rich counterparts. This means that there must be a conspiracy between mass-to-light ratio and stellar surface density of the disk. Gravitational equilibrium requires that the former should increase as the latter is reduced. One explanation attributes this connection to a variation in star formation efficiency, which could be driven by supernova feedback, although no detailed model has been developed.

Elliptical Galaxies

The bright galaxy end of the luminosity distribution of galaxies is dominated by early-type galaxies and displays an exponential cutoff with a characteristic luminosity of 30 billion solar luminosities. In the canonical cold dark matter cosmological model, gas infall continues until the present epoch. Approximately 90 percent of the baryons are presently in the intergalactic medium. These baryons constitute a huge gas reservoir that potentially is available for accretion onto galaxies and eventual star formation. However, gas cooling and significant amounts of star formation are not observed for massive galaxies.

Most are red and dead, and essentially all star formation terminated a Hubble time ago. More generally, it is believed that massive spheroid formation is terminated by the effects of active galaxy outflows on the halo gas reservoir. Much of what comes in must go back out. Outflows from active galaxies quench infall, strip the halo of gas, and lead to old (and red) ellipticals.

There is a similar gas cooling problem in galaxy clusters. Cluster galaxies are predominantly ellipticals, especially in their centers. But clusters are full of hot intergalactic gas that inevitably cools in their centers. Gravitational collapse initially keeps the gas hot, but once the cluster forms, this heat supply is interrupted. Gas inevitably cools down once the heat source is turned off. Hence, according to theory, the gas should fall into clusters that become galaxies, which in turn form stars and turn blue. But we do not see this happen. Something is providing late heating—and stopping the cooling down.

If late heating is due in part to gravity, cold infall into galaxies could possibly never happen. Gravitational infall results in gas heating. So does the associated drag. These effects may suffice to explain late quenching of star formation by suppressing the cold gas supply. In contrast, active galaxy outflows provide early as well as late suppression of star formation. There is another way to explain late-quenching: we can ask whether the many low-luminosity active galaxies we see might have inhibited cooling flows in galaxy clusters. At the current epoch, there is still substantial gas accretion. Somehow one has to stop the gas from forming stars in great numbers. Low-luminosity active galaxies may heat the gas but cannot eject it via winds. Hence, one expects the halos of massive galaxies to have a mass of diffuse hot gas comparable to the stellar mass.

From an observational point of view, our studies of massive ellipticals show that there is a baryonic shortfall within the halos of these galaxies. However, the baryons are certainly present on group and cluster scales, in the form of intracluster stars and diffuse gas. Moreover the gas is enriched, which provides an important piece of evidence about its history. Enrichment of the gas by supernovae must have occurred.

From this circumstantial clue we realize this: presumably the gas was ejected by intense supernova input, most likely in the early stages of galaxy formation. In the case of the ellipticals, active galaxy outflows can provide the energy source, but these are inevitably associated with vigorous star formation activity. Observations of ultraluminous infrared galaxies, the most luminous galaxies in the universe, show evidence for winds. In these galaxies, the bulk of the stellar mass has already been assembled. Nevertheless, they have luminosities in excess of one hundred Milky Way galaxies.

Corresponding star formation rates are as high as a thousand solar masses per year. When we see such phenomena, it is likely we are witnessing the final phases of ellipticals in formation. Indeed, strong winds are essentially ubiquitous in ultraluminous galaxies. The outflow rate is usually comparable to the star formation rate, which suggests that indeed the mass of ejected baryons is comparable to that remaining in stars.

What is less clear is whether the high efficiency needed for massive spheroid formation, already in place by a redshift of 2, is due to the role of active galaxies in triggering star formation. Star formation timescales, as short as 10 million years, are inferred for massive elliptical galaxies. (Star formation timescales are determined by making a synthesis of the different types of stars needed to account for the detailed spectrum of the galaxy.) We can infer from the colors in effect if the galaxy is young or old.

One clue to formation is that active galactic cores are present in a large fraction of the total number of ultraluminous galaxies. Unfortunately there is a chicken-and-egg type of problem here. The overlap does not tell us whether the active nuclei are a trigger or a consequence of the ultrahigh rates of star formation. Either way, something is needed to accelerate star formation, and supermassive black holes are a likely ingredient. Active nuclei plausibly play a dual role in spheroid formation, both in triggering and amplifying as well as in quenching star formation.

Feedback

A strong case can be made that feedback, the exchange of material between different developing objects, plays an important role in galaxy formation. Unfortunately, we do not have enough robust information to simulate in computer models the exact feedback mechanisms that are at play when galaxies form. Our observations, however, show us three kinds of activity: mergers of galaxies, the dynamics of supermassive black holes, and supernova turbulence.

When the merger of two or more galaxies takes place, it drives gas into the center of the more massive galaxy. One finds evidence for merging in many ultraluminous galaxies. The spheroid star formation mode is plausibly motivated by gas-rich mergers. However, there are not enough major mergers especially at high redshift to account for all ultraluminous galaxies. Also, gas-rich mergers form ellipticals with certain morphological characteristics that are reminiscent of embedded disks because the gas acquires angular momentum, some of which is retained by the newly formed stars. We simply do not see this effect in the most massive ellipticals. They could not have formed this way. It is likely that minor mergers are the dominant gas supply mode. The gas supply also helps feed the central supermassive black hole.

Supernovae cannot account for outflows from massive galaxies. The nuclear outflows play an essential role in quenching star formation; otherwise too many ellipticals are too luminous and too blue. Active nucleus-driven outflows are a natural consequence. Supermassive black hole growth is in phase with galaxy formation, or at least with the global star formation rate.

Our second example of feedback comes from supermassive black holes. They are fed during the gas-rich phase of galaxy formation. Correlation of supermassive black hole mass with the spheroid potential suggests that formation of both is contemporaneous. The starburst association with spheroid formation occurs during the supermassive black hole growth phase. Active galactic outflows can also trigger star formation before quenching it. The phenomena are most likely causally connected, but we do not know the associated arrow of time. That is, does supermassive black hole growth precede and trigger spheroid formation, or is supermassive black hole growth a consequence and eventual terminator of spheroid formation? In the gas-rich formation phase, the momentum input from active galactic nuclei is dominant by an order of magnitude or more relative to input from supernovae. Indeed, supernovae provide negative feedback in galactic disks. They drive winds from potential wells of low-mass galaxies. Supermassive black holes induce outflows that provide positive feedback in massive protospheroids. Blowout occurs and star formation terminates when the supermassive black hole reaches a critical mass that depends only on the spheroid mass. The supermassive black hole–quenching hypothesis leads to a supermassive black hole–spheroid velocity dispersion correlation that fits the observed relation between black hole and spheroid mass, both in absolute value and in slope.

Finally, the great explosions of supernovae and other active galaxy outflows play essential roles in how feedback leads to galaxy formation. In disk galaxies, supernova remnants self-regulate the interstellar medium, driving fountains out of the disk. The gas cools in the halo and eventually accretes onto the disk as high-velocity clouds. During spheroid formation, the central supermassive black hole growth phase results in outflows that induce accelerated star formation. The induced supernovae can drive winds even at early stages of spheroid formation. Only when the supermassive black hole attains its final mass is the active galaxy wind strong enough to clean out the spheroid and thereby terminate any subsequent star formation.

As noted, we do not have a robust mechanism in computer simulations to explain feedback. The situation is especially grave in not having data to explain the forming of spheroids in general. Here, one lacks the framework of the overall disk gravitational instability that sets the scene for star formation in disk galaxies. In disk galaxies, for example, the instability drives spiral density waves that in turn generate molecular cloud complexes. The subsequent steps involve cloud fragmentation and turbulence generation. This process culminates when the cloud core collapses and stars are formed.

Our understanding of these successive steps involves as much phenomenology (namely, our observations) as theory. Indeed, our modeling results depend entirely on the phenomenological data. For example, while supersonic turbulence is observed to dominate on all scales of the interstellar medium, what drives the turbulence is disputed among theorists. Large-scale gravitational instabilities, magnetically driven instabilities, supernova explosions, and protostellar outflows are all conjectured to play a role.

Most likely, all of these dramatic events in space provide input at some level to turbulence over a wide variety of scales. The output should be insensitive to the details of the input. If not, our models would be in difficulty. Perhaps the most serious issue is that whatever mix of physics is relevant to the feedback events that we see in the nearby universe, there is little reason to be confident that a similar prescription holds at early epochs in the universe. All predictions for the high redshift universe (that is, the early universe) should be regarded as, at best, inspired guesswork.

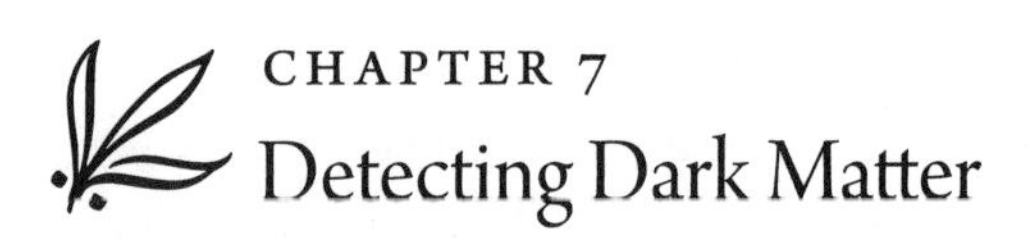

CHAPTER 7
Detecting Dark Matter

IDENTIFYING DARK MATTER is one of the greatest challenges in cosmology. Ninety percent of the matter in the universe is dark, but we can only speculate about its nature. The leading candidate for dark matter is some variety of very weakly interacting elementary particle. This candidate must interact with the material in the universe in a way that is much weaker than the proton, a basic constituent of all matter.

This weak interaction of dark matter explains its apparent behavior in the universe. When a cloud of dark matter and ordinary matter contracts to form a galaxy, for example, the ordinary matter eventually fragments into stars, but the dark matter is left behind. Its weak interaction also means that dark matter particles do not lose energy by radiation, and hence their "darkness." During cosmic evolution, dark particles condense into dark clouds, but not into stars. Dark matter is pulled by gravity alone.

In our observations today, we find dark matter both in the inner regions and outer regions of galaxies. In our own Milky Way galaxy, for instance, there is an increasing concentration of dark matter in the inner galaxy, in contrast to the way ordinary matter dominates in regions that include our solar system, which is ordinary matter orbiting our sun. Dark matter also populates the outer regions of galaxies, where there are fewer stars. Dark matter dominates the mean density of these outer regions. Nevertheless, these stars, far from the galaxy center, continue to maintain their orbital velocity. If the distribution of the dark matter were traced by the stars, the circular velocities would decrease as the gravity field gets weaker. The dark matter is much less concentrated than the starlight.

Dark matter consists of weakly interacting elementary particles. We know that any particle that can exist according to the laws of physics once did exist in the fiery furnace of the big bang. All particles have antiparticles of the opposite charge. Today antiparticles are exceedingly rare, but in the first instants of the big bang every particle coexisted with its antiparticle. There was complete symmetry. This changed, however, as the universe cooled down.

The primordial furnace left behind very little antimatter, which is fortunate for us because an encounter between a particle and an antiparticle results in total annihilation. No matter survives. The amount of surviving antimatter depends only on the strength of the interactions of the particle in question. An antiproton is strongly interacting. Many were once present. Pairs of protons and antiprotons were created spontaneously from radiation at the extreme temperatures encountered in the first nanoseconds of the universe. In fact, the universe was once dominated by proton-antiproton pairs. As the universe expanded and cooled, spontaneous creation stopped. Almost all of the antiprotons annihilated with protons (but fortunately for us, some protons are left behind to form the material universe). This great annihilation produced the cosmic microwave background we measure today, the heat left over from the fireball at the beginning of the universe. The same logic applies to the dark matter particles. Most had annihilated. But in this case far more were left behind because these particles interact so weakly.

At this point in our exploration of the dark side of the universe, we discovered a remarkable coincidence in how the forces of nature seem to work. This kind of discovery always makes cosmologists feel good inside. Americans Ben Lee and physics Nobel laureate Steven Weinberg discovered that the leftover abundance of weakly interacting particles was just what is needed to account for the dark matter. There was no reason whatsoever for this coincidence:

nature just prescribes it. Weakly interacting particles can actually account for all of the observed dark matter if their interaction strength is typical of the weak nuclear force. They must, of course, be long-lived and interact with a strength similar to that of the neutrino. This seems a natural interaction strength to be expected for such particles.

The explanation of the origin of the weakly interacting dark matter soon became even more compelling. A new theory, called supersymmetry, was developed that postulated the existence of hundreds of possible dark matter candidates. (Recall from an earlier chapter that symmetry was the point at which the universe was so hot that nothing was separated, as when we melted a Rolls-Royce and Volkswagen together so their basic chemicals were the same.) In the new supersymmetry theory, any candidate for dark matter must be a weakly interacting particle. Only one of the candidates can be the real thing, however, and in fact it must be the lightest of the supersymmetric particles so that it is sufficiently stable. The rest of the candidates live for only nanoseconds.

We are now hot on the trail of these particles. One of the primary goals of the Large Hadron Collider, operated by CERN on the Franco-Swiss frontier, is to search for evidence of supersymmetry. The circular accelerator, which runs for seventeen miles, will seek one-sided jets in high-energy proton-antiproton collisions. The collision releases energy that in turn creates cascades of quarks and other short-lived particles. Momentum must be conserved in the collision. Seeing a jet of particles emerge in one direction with no corresponding jet spraying out in the opposite direction would provide evidence for supersymmetry.

As noted earlier, there are many candidates for the dark matter. Fortunately, the most popular class of candidate particles hails from the theory of supersymmetry and is likely to be observable, provided that supersymmetry is correct. The lightest supersymmetric particle is expected to be stable and can account for the dark matter density. Supersymmetry is motivated by theory but also by empirical arguments. One of these is the running of the coupling constants of the fundamental forces. The constants are very different from each other today, but this was not always the case. Today we live in a low-energy universe, but at very high energies the weak nuclear force becomes much stronger. When we extrapolate the strengths of the different forces measured at low (100 GeV) energies, they seem to converge near 1,000 trillion GeV. Supersymmetry requires convergence at a precise energy, which is called the scale of grand unification of the fundamental forces. As we saw earlier, this scale provided the stimulus and natural setting for inflation.

Again, the candidate particle must be neutral and weakly interacting. It is referred to as the neutralino and is the heavy counterpart of a known particle such as the photon. Neutralinos are in thermal equilibrium at the high temperatures of the very early universe. As the temperature dropped at the beginning of the universe, a relic density of all the stable particles survived. The weak nature of neutralino self-interactions resulted in a large relic density, in contrast to that resulting from strongly interacting particles. Remarkably, the relic density generically gives an approximately critical density of dark matter for a weaklike cross-section.

The universe was once at energies where supersymmetry prevailed and hence a breeding ground for supersymmetric dark matter. This means matter and antimatter. The primordial furnace must have created pairs of weakly interacting particles and their antiparticle partners. Particle theory gives us candidates for these particles. None have actually yet been observed, which is a minor obstacle. If one of these new particles is stable, then its weak interactions guarantee a large survival fraction today. These could be the elusive dark matter. Now we can understand why there were so many dark matter particles left over from the big bang.

We are trying to find this dark particle by the use of massive machines using state-of-the-art technologies like the Large Hadron Collider. But this project alone cost some 5 billion euros.

The collider has put together thousands of superconducting magnets in order to produce a particle accelerator that, in its circular ground plan, is five miles in diameter. Obviously, a machine like this can become so expensive that we reach the earthly limit for such research. Fortunately, therefore, we can still use the universe itself as a particle accelerator. In the reaches of the universe, particles are annihilating and stream across space. We are bombarded with dark matter. About 10 million particles per square meter per second pass through us. As we will soon see, we don't always have to build a supercollider: the universe itself provides a poor man's way of searching for dark matter.

The dark matter that bombards us is produced by the annihilation of particles (as if crashing in a supercollider) in the distant universe. How is this taking place? Once, in the very early universe, the dark matter particles were far more numerous, and annihilations were frequent. Today, the density is low, and annihilations of the dark matter particles are very rare. But although rare, because the volume of each galaxy halo is so large, annihilations may still lead to a detectable signal. Hence, our best target is the debris from particle annihilations. Typical particle masses are expected to be on the order of the energy scale characterized by our current theory of supersymmetry. This scale is between one hundred and one thousand proton masses. The acronym for these weakly interacting massive particles is WIMPs.

When annihilations take place, they produce a shower of energetic elementary particles, most of which decay immediately. The stable products are high-energy gamma rays, proton-antiproton pairs, neutrinos, and electron-positron pairs. All of these are potentially observable with future experiments. The predicted flux depends on the density of dark matter. We infer the local density in the solar neighborhood from the rotation of our galaxy. It amounts to a third of a GeV per cubic centimeter. Some 10 million per square meter per second pass through the Earth. Nothing beats direct detection. This enormous flux of weakly interacting particles illustrates the nature of the research challenge for the future.

With such a great influx, we can try to achieve a direct detection of the particles. In direct detection techniques, the dark matter particles, although very weakly interacting, do occasionally collide with atoms in a device—the detector—that has a special highly sensitive material to pick up a slight trace from particle collisions. But this direct detection does not work with other kinds of dark matter annihilation, so we use indirect detection as well. This technique picks up the occasional dark matter annihilations in a galaxy halo. When these annihilations occur, they produce an energy loss in the form of both gamma rays and energetic particles. Our detection of these is indirect confirmation of dark matter.

Both kinds of detection are giving us hope to solve the dark matter puzzle. It's also a fascinating story of technique and technology. So let's look more closely at both of these approaches, beginning with direct detection.

Direct Detection

Clearly, there is no substitute for direct detection of the dark matter particles. The flux of such particles at the earth is relatively large. Of course the interactions are weak, but elastic recoils leave potentially detectable signals in sufficiently large masses of target material. The nuclear recoils heat and occasionally ionize detector nuclei. Detectable signatures include seeing traces of electrons and ions as well as pulses of vibrational energy that occur in the atomic lattice structures of all solid materials called phonons. However, there are important backgrounds, especially resulting from cosmic rays hitting the detector and producing neutrons. To see the WIMP signal, one has to suppress these backgrounds. The only way is to go deep underground. Disused areas in mines have been used. One of the most effective is in Sudbury, Canada, at a depth of three kilometers.

Until recently, detectors in underground laboratories have been kilogram-scale, with sensitivities to elastic scattering cross-sections down to a ten-billionth of a barn (or 10^{-24} cm^2). One barn is the typical interaction cross-section for ordinary atomic nuclei such as protons. It corresponds to an effective area for scattering amounting to the square of the classical radius of a proton. These are the strong nuclear interactions. For there to be a significant density of WIMPs surviving from the very early universe, the typical interaction strength must be less than a billionth of a barn. This is the domain of the weak nuclear interactions. These are the reactions that involve neutrinos. The experiments have ruled out WIMPs in this regime.

However, the supersymmetric model parameters allow a significant density of dark matter over a wide range of interaction strengths, below those of normal weak interactions. One can still have WIMP dark matter for cross-sections up to three orders of magnitude lower than current experimental limits from WIMP searches. Theory has shifted the goalposts and more sensitive experiments are required. The only way forward is to have larger mass detectors. Ton or even kiloton detector masses are required to probe the full range. This is the goal of the new generation of experiments that use detector materials as varied as liquid argon and xenon or crystalline germanium.

One intriguing development has come from the realization that the dark matter may possess fine-scale structure on the scale of the solar system. The dark matter is cold, in the sense that individual particles have essentially no random motions. If fluctuations are present initially, the dark matter responds to gravity on scales much smaller than those of galaxies. Indeed the smallest clumps expected have earth masses but are much more diffuse than a planet. Consequently, the earth as it orbits the sun occasionally runs into dark matter clumps. The dark matter is so diffuse that it has no direct influence via impact on the earth. However, experiments searching for the weak signal from direct detection would find an annual modulation of the elusive signal. Streams of dark matter in the solar neighborhood result from the incomplete tidal disruption of dark matter clumps in galactic orbits. One direct detection experiment at Gran Sasso, a deep underground laboratory in Italy, makes use of this signature and is the only experiment to report a claimed detection. This experiment, called LIBRA, utilizes a vast vat of sodium iodide and searches for light scintillations. Other groups dispute their results, however, using different techniques and more sensitive experiments.

A window remains, however, if the neutralinos interact coherently. In this case, the interactions depend on the direction of spin of the target atomic nucleus, which can be important for nuclei with odd numbers of particles. Very few experiments are sensitive to this interaction. The most important case is hydrogen. The sun can trap such WIMPs, and its structure is very slightly affected. If the mass is around five proton masses, the WIMPs take a long time to settle into the core and annihilate. They basically fill the sun. The usual lower bound on WIMP mass is about fifty proton masses from accelerator limits. But some models permit these lower masses. Low-mass WIMPs turn out to be a possible means of reconciling LIBRA with most of the other experiments. These WIMPs scatter with solar particles and slightly modify the temperature profile by redistributing the thermal energy.

There are two techniques for measuring the internal temperature in the sun. One involves detecting the neutrinos emitted in the thermonuclear reactions that occur in the solar core. The successful detection of solar neutrinos has confirmed that thermonuclear burning of hydrogen powers the sun. It also fixes the central temperature to high precision. Another experiment involves detecting seismic quakes in the sun. Helioseismology measures very slight vibrations of the sun. The frequencies of the measured modes depend on the sound crossing time in the sun, and therefore on the temperature profile. At present, our best model of the sun cannot account for the helioseismology results. The discrepancy is only about 10 percent. However, the conclusion is that despite the sun being the closest star and exquisitely well studied, we still lack a complete understanding of its structure. Dark matter may conceivably have a role to play, although it is far more likely that the discrepancy lies in something as mundane as our modeling of turbulence inside the sun.

Indirect Detection

As we described earlier, our indirect detection of dark matter is achieved by tracing the pieces that spray off other particles when they collide with dark matter. When dark matter particles of one hundred to one thousand proton masses collide with other particles, the annihilation produces energetic gamma rays, neutrinos, positrons, and antiprotons. All of these are distinct signatures that can be observed against a known background. High-energy positrons that emerge from collisions, for example, are especially useful since they are a rare component of the galactic cosmic rays. Moreover, all of these by-products of the collision look different from relics of the dark matter, since those relics cannot exceed the WIMP mass. This contrast leads to a feature in the energy distribution of the high-energy photons or particles. Other particle backgrounds are blind to WIMPs, and so the backgrounds do not show any features that might indicate the presence of WIMPs.

Despite the many candidates for the lightest supersymmetric particle, we know its annihilation cross-section to great accuracy. This stems from the requirement that the WIMPs account for the density of dark matter. If the cross-section is too low, there are too many WIMPs. If it is too high, there are too few WIMPs. This constraint arises because annihilations in the early universe fix the surviving density of WIMPs.

The WIMP particle so far is undetected, which, of course, means that its mass is unknown. However, the supersymmetry theory specifies a range of cross-sections at any given mass. This range spans some three factors of ten, while the allowable mass range

Series A: *Views of Star Birth*

The star-forming region, NGC 3324, in the Milky Way

The Orion nebula, a combined image from the Hubble space telescope and the Spitzer infrared space telescope

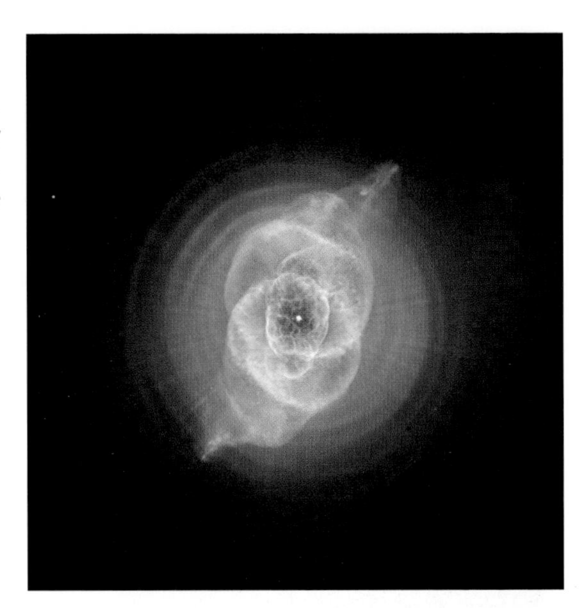

NGC 6543, a planetary nebula known as the Cat's Eye

The Tycho supernova remnant, the residual debris of a white dwarf star that exploded in 1572

SERIES C: ***Star-forming Galaxies***

The spiral galaxy, Messier 51, along with a small companion galaxy

The barred spiral galaxy, NGC 1672

The Sombrero galaxy, seen in a combination of infrared and optical light

Our neighbor, Messier 31, the Andromeda galaxy

Above:
The globular star cluster, Omega Centaurus, NGC 5139

Left: A giant elliptical galaxy, NGC 4649, viewed with the Chanda X-ray satellite

An active galaxy, Messier 82, currently undergoing a violent galactic-scale explosion

Two galaxies undergoing a merger, spiral NGC 2207 and IC 2163, viewed with the Hubble and Spitzer space telescopes

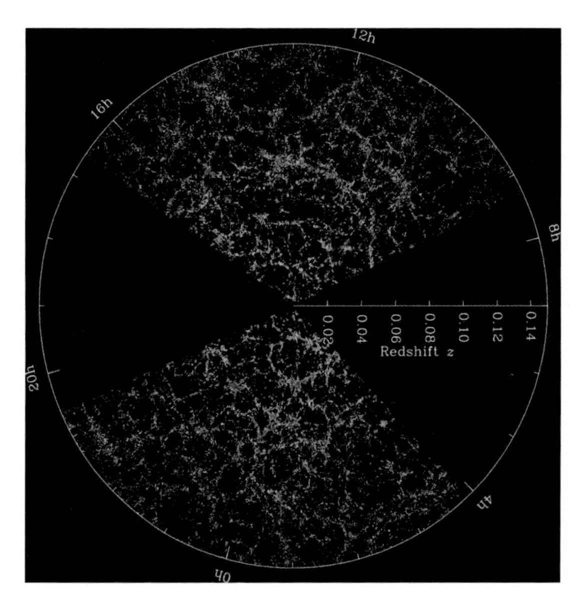

The Sloan Digital Sky Survey, a map of nearly a million galaxies

The cosmic microwave background: five years of data from the Wilkinson Microwave Anisotropy Probe satellite

SERIES G: ***Birth of the Universe and the Clustering of Galaxies (continued)***

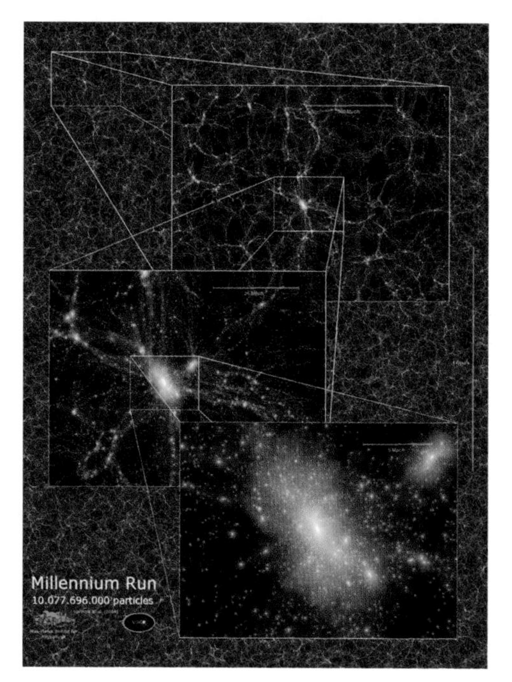

Right: The Millennium numerical simulation of 10 billion particles—formation of a galaxy cluster

Below: Virtual galaxy surveys extracted from the Millennium numerical simulation, compared to the actual data

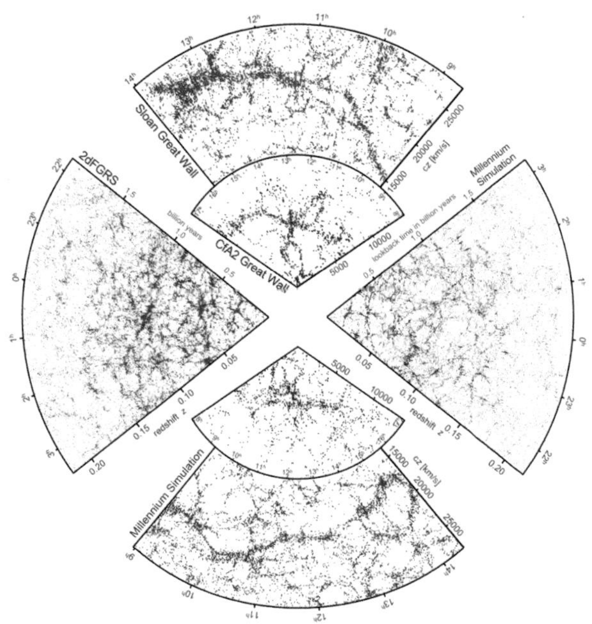

extends from only 50 GeV to 1 TeV. The inference is that experiments must cover a wide range of sensitivities to have any chance of detection. However, detection at some level is guaranteed if the lightest supersymmetric particle is the partner of one of the known particles, which themselves are classified according to their spin and mass into particles such as bosons (for example protons) or fermions (which include electrons and quarks).

The positrons provide a promising signal because cosmic ray positrons are quite rare. The positron flux is controlled by the local dark matter density. The gamma ray flux is also an interesting signal. The halo is transparent to gamma rays. Because the density of dark matter is higher toward the center of the galaxy, the gamma ray flux is boosted. There are gamma rays produced by the interaction between cosmic rays and interstellar gas. There are also positrons in the galactic cosmic rays. But the dark matter signal should stand out. Its positrons and gamma rays have a unique distribution of energy, determined by the mass of the annihilating particle. Detection of excesses in both the positrons and gamma rays relative to the known backgrounds would be smoking-gun evidence of dark matter particles.

Experiments carried out far above the surface of the earth have been performed to look for positron and gamma ray evidence. It is necessary to go into space because there is too much contamination by cosmic rays interacting with the earth's atmosphere. So far, there is no definitive indication of a signal in either gamma rays or positrons that could be from dark matter.

There are, however, hints of possible detections in both cosmic ray positrons and diffuse gamma rays. For example, high-flying balloon experiments reported an anomalous spectral bump feature in the positron flux that is not explained by the featureless secondary spectrum of positrons generated by cosmic ray interactions with interstellar matter. The dark matter particles annihilate into a shower of debris. Consequently the annihilations yield a positron feature centered around 10 percent of the particle mass. However, the data was not very convincing. The uncertainties are large. It is too premature to make much of any possible annihilation signal. New experiments are under way.

A new experiment on an Italo-Russian space satellite launched in 2006 called PAMELA is taking data that after several years in orbit is providing a much-improved measurement of the positron flux. Early results support the anomalous spectral bump, but the interpretation of this feature cannot necessarily be interpreted as a dark matter signal. The strength is much larger than expected from the simplest theories. The alternative explanation appeals to a nearby pulsar in the constellation of Gemini called Geminga (an abbreviation for Gemini gamma ray source) that is known to be a source of very high-energy gamma rays. More data will be needed to better understand the experimental results, and it will be necessary to find complementary signals. The most promising of these comes from diffuse cosmic gamma rays.

Gamma rays from space have also been detected. Indeed the Milky Way glows in gamma rays that are produced as cosmic rays pass through interstellar gas clouds. High-energy collisions between protons generate gamma rays. The COMPTON gamma ray observatory detected an excess diffuse flux of gamma rays from the innermost galaxy. The energies of these gamma rays seemed to be higher than expected from the flux predicted by cosmic ray interactions with interstellar gas. More exotic decay channels may be indicated than would be achievable with cosmic ray protons.

This is just what is predicted for the WIMPs. The available decay channels for WIMP neutralino annihilation involve massive quarks, which weigh up to five proton masses. The resulting gamma rays from decays of these quarks are more energetic than those produced when protons collide with other protons. However, the data need to be confirmed. The Gamma Ray Large Area Space Telescope (named FERMI after Italian American physicist Enrico Fermi) was launched in May 2008 and should provide the necessary confirmation if WIMPs are indeed the elusive dark matter. The FERMI telescope provides an improvement in sensitivity by a factor of one hundred over the previous gamma ray satellite experiment, flown on the Compton Gamma Ray Observatory. This experiment will greatly improve on the existing data. It should be possible to confirm (or reject) the reality of the possible signal from annihilating WIMPs.

If the preliminary early results were to be interpreted as due to WIMP annihilations, a serious issue arises concerning the strength of the predicted gamma ray and positron signals. What has been (possibly) observed is far larger than expected. A uniform dark halo falls short by nearly a factor of a hundred in accounting for the observed fluxes in terms of plausible annihilation rates from the known dark matter density. There is a resolution of this dilemma, at least for the gamma rays.

Clumpiness and Spikes

In our search for dark matter, cosmologists must also contend with two other kinds of phenomenon, which can be helpful in our detection of the mysterious particle. One of these is the clumpiness of the dark matter, and a second is the way that dark matter spikes when it is in the vicinity of a black hole.

The clumpiness of dark matter is related to its ability to provide the earliest pattern for the observed large-scale structure of the universe, which develops later. In the very early history of the universe, the WIMP particles decoupled from the radiation, which means that in the recent past, the WIMP particles were completely cold (that is, they had negligible random velocities remaining from their thermal origin). As the universe expanded, random motions of particles would decrease and the WIMPs would become cold. This is the connection between cold dark matter and large-scale structure. The detailed clustering properties can be reproduced in computer simulations, if simple models for galaxy formation are used. However, the cold dark matter candidate has not yet been detected: it remains a hypothesis.

One uncertainty in computing the flux is the degree of clumpiness of the dark matter. For now, we work on the assumption that the annihilation rate is proportional to the square of the density. Numerical simulations of galaxy formation show that halos are clumpy, with up to 10 percent of the dark matter being in clumps, which can result in a large boost factor for the annihilation flux. The theoretical modeling of the halo tells us that it is highly clumpy. In a clumpy halo, there are far more particle collisions and annihilations than in a smooth halo. It is quite easy to account for the observed gamma ray flux in this way.

The positrons cannot be so easily explained because positrons scatter frequently as they propagate in the interstellar medium. The observed positrons cannot come from very far away, and certainly not from the inner galaxy where the enhanced density and clumpiness of the dark matter boost the observed flux. The positron story must be more complicated, depending on how the positrons propagate and accumulate.

The flux of the annihilation signal is sensitive to the minimum clump size—about the size of an earth mass for typical WIMPs. However, interactions with stars will tidally disrupt many of the clumps, effectively depending on their orbits. Clumps with orbits highly inclined to the disk should survive. Low-mass clumps are only visible via their annihilation signal, which is obtainable by the indirect detection method.

We have only mentioned black holes briefly so far, but now they become an important factor in detecting dark matter as well. Dark matter is concentrated in the vicinity of a black hole. Furthermore, relic spikes of dark matter develop around forming black holes. Dark matter particles are enhanced in number in the locally increasing gravity field. The dark matter density is increased near the black hole. So is the annihilation signal.

We first need an overview of black hole formation. Their formation requires matter that can lose energy as it collapses. Baryons can do this by radiative cooling, but dark matter cannot, so it does not become material for the black hole. But, as the baryons aggregate to form the black hole, some dark matter particles invariably are trapped by the increasing strength of the local gravity field. The annihilation flux may be greatly boosted near a massive black hole that forms in the nucleus of a forming galaxy. These black holes are expected to form by way of accretion of baryons onto a seed black hole—one that is just sprouting, so to speak. The seed is formed from the debris of massive stars. This is not the only formation possibility, and another route to massive black hole formation is via runaway mergers of stars. We observe indirectly the phase of black hole growth as the consequences of gas accretion. The luminosity of the black hole increases as it is refueled by infalling debris. We infer the amount of fueling since this is the phenomenon that powers quasars and active galactic nuclei.

We have other clues about how black holes form and their relationship to dark matter. One is the fact that supermassive black holes usually show up in the spheroid, or large bulge, at the centers of galaxies. This seems to be a universal phenomenon. A substantial fraction of the mass of a spheroid is dark matter. When the central black hole of a spheroid formed, it must have concentrated the local dark matter in its vicinity. The weakly interacting dark matter responded to the deepening gravitational potential well around the black hole. A density spike of dark matter developed within the gravitational sphere of influence of the black hole.

In looking for dark matter spikes, we naturally recognize that black holes form with a wide range of masses. The smallest dark matter clouds in which baryons can cool amount to about a million solar masses in weight. The most massive correspond to the largest galaxies. The corresponding black hole masses range from a thousand solar masses to a few billion solar masses. All of these black holes develop dark matter spikes or cusps.

Black hole mergers destroy these nearby cusps of dark matter. Hence, it is unclear, for instance, whether the 3-million-solar-mass black hole at the center of the Milky Way possesses such a cusp. If this were the case, the annihilation rate would be greatly enhanced in the vicinity of the central black hole, and this annihilation rate would be detectable via gamma ray emission. But not all black holes move to the center of galaxies. For instance, during the hierarchical buildup of the Milky Way, which involved merging together clouds of dark matter and baryons, the smaller black holes did not merge into the central black hole. These intermediate-mass black holes remain in orbit in the inner galaxy. They are ubiquitous, and they are potential targets to detect dark matter spikes.

Black holes are frequently ejected from the centers of galaxies, perhaps into intergalactic space. A merger of two black holes often ends with a one-sided burst of gravitational radiation. The net result is that the merged black hole receives the equivalent of a kick in the opposite direction, since momentum is conserved. The acquired velocity is often large enough to eject the black hole from its host galaxy. These intergalactic black holes should still retain some dark matter in their vicinity, although merging tends to weaken any initial spike of dark matter.

The spikes should survive around intermediate-mass black holes that have not undergone merging. These thousand-solar-mass black holes are relics of the initial clouds within the first stars that formed. The most massive of these clouds would have cooled as the emitted radiation from hydrogen atoms that are excited by collisions with other atoms. This so-called Lyman alpha emission, and the efficient cooling associated with this type of cooling (which requires only atomic hydrogen), would have facilitated gas accretion rates onto forming protostars. The enhanced accretion implies that the first stars were necessarily massive. Massive star collapse facilitates the likely formation of thousand-solar-mass black holes.

The most massive black holes in the universe power up ultraluminous quasars. These are the biggest black holes, and their masses exceed a few billion solar masses. They are found in the early universe, at a redshift of 6 or beyond. These black holes are so massive and in place so early that intermediate-mass precursor black holes must have existed in order to explain their formation. These intermediate-mass black holes could be the missing link between primordial clouds, primitive early states, and quasars (which form later around the largest black holes).

As noted, most of these smaller black holes failed to merge into the larger ones, and so are leftovers. In general, the merging process is notoriously inefficient anyway. Many smaller black holes would be ejected by recoil in the form of binaries (which are black holes in orbit about each other), and then simply be dissolved as the larger systems grow. But as many leftover black holes survive as are dissolved, and these are the relic black holes that we believe populate our halo. Added up, these intermediate-mass black holes are expected to account for as much mass as the central supermassive black hole in a galaxy. Most important for our purposes here, these intermediate black holes retain their initial dark matter spikes. These black holes are primary candidates to study for gamma rays and neutrinos that are produced by annihilations of the dark matter, now in the form of relic spikes.

And if finding and detecting dark matter is not challenging enough, cosmologists today also are grappling with dark energy, an even more unexpected discovery.

CHAPTER 8
Finding Dark Energy

THE UNIVERSE is walking a fine line. According to our measurements, the universe is very, very close to being flat. The total mass of the universe keeps it at the critical point between being open—that is, expanding without limits—and closed, which means its space is curved and it will collapse back on itself. In a closed universe, gravity overcomes the expansion. But in an open one, the outward movement is stronger than gravity. What we find is that the universe is in a kind of balance between open and closed, even though, as we mentioned earlier, its flat space is expanding at an accelerated rate. The universe may be very slightly curved, in an open or closed sense. This will turn out to be of interest when we discuss below whether the universe is infinite. What, then, is keeping the universe in balance so close to flatness, and what is causing the accelerated expansion? The answer may be dark energy.

For a start, the visible matter that we find in the universe is only a third of the amount necessary to keep the universe in a balance between gravity and expansion. In a sequel to his early theory of general relativity, Einstein solved the imbalance by proposing that a cosmological constant, or force, counteracts gravity. This gave Einstein a static universe, a model that has now been discarded. As we saw in an earlier chapter, cosmologists Georges Lemaître and Arthur Eddington revived the idea of the cosmological constant as the force that explains the expanding universe, and our current model is based on the Friedmann-Lemaître theory of the big bang.

Today, the modern interpretation of the cosmological constant is that it is energy with a negative pressure, a repulsive force that pushes outward. In contrast, positive pressure acts like gravity and exerts an attractive force, and this attraction curtails expansion of the universe. The repulsive force of negative pressure is like the tension in a string. It acts as a constant contribution to the energy density of the universe. If dark energy accounts for two-thirds of the universe's critical density (which we can't find in the matter itself), the flatness of the universe is explained.

This theory has a remarkable consequence for how we picture the history of the universe. All energy densities decrease bar one, that of dark energy, which is just the cosmological constant. At early epochs, the energy density of the cosmological constant would have been dominated by the density of ordinary matter and radiation. Its effect was negligible. However, at late epochs, the dark energy began to dominate. When the dark energy density is larger than the redshifting matter density, the universe enters into a phase of acceleration. This effect has been detected, thereby confirming the dominance of the dark energy.

The modern Hubble diagram revealed a phenomenon that would have astonished Hubble. Supernovae in the most distant galaxies are found to be unexpectedly dim. The best explanation advanced to date is that the expansion of the universe is accelerating. The supernovae are farther away than would be the case in a nonaccelerating universe.

There are bizarre consequences. The future dominance of acceleration means that the number of visible galaxies is decreasing. Currently we can see some 10 billion within our horizon. Someday in the far future our Milky Way will be alone.

Chasing Supernovae

To understand how our universe may be accelerating by the force of dark energy, we need to measure distances of galaxies. For this, supernova stellar explosions are a key. As we saw earlier, supernovae are the events that seed the universe with heavy elements, allowing the evolution of stars and galaxies. But the brightness of their explosion also has turned out to be a remarkable tool for measuring distances. Fortunately, about half of all supernovae have the characteristics that allow for these measurements.

Type Ia supernovae have characteristic spectra, dominated by iron lines, and are also more luminous than Type II supernovae. The latter are associated with the deaths of massive stars. Type Ia supernovae explode when white dwarfs, the relics of low-mass stars, accrete mass from, or even merge with, a companion and become unstable. Collapse occurs when the white dwarf is driven above the Chandrasekhar mass of about seven-tenths of a solar mass. The energy released in the collapse is the source of the explosion. The explosion energy comes from the nuclear energy available when the carbon and oxygen of the white dwarf is burned into iron. The amount of nuclear energy is 10^{51} ergs, which is converted into light by the radioactive decay of nickel which turns into iron. Thus, we have a universal light source. A supernova at maximum light shines with the brilliance of a billion suns, brighter than many galaxies. Supernovae can be detected far away at a redshift of 1 or beyond, and provide the most powerful means of measuring distances to remote galaxies. Our only assumption is that a remote supernova of Type Ia shines with the same brilliance as a nearby one. Because its luminosity is determined by fundamental physics, this assumption is well founded. In this way, supernovae were used to extend Hubble's diagram that plotted recession velocity of a distant galaxy against its distance.

In 2000, cosmology was changed by a breakthrough in our understanding of how to best use supernovae to measure distances. Until then, supernovae had been our best method, but the accuracy of distance scales was relatively poor. There is an intrinsic range in supernova brightnesses, and without understanding this range completely, it was easy to arrive at large uncertainties.

Then in 2000 came the new discovery: Supernovae have a characteristic light curve, an exact rise and fall of luminosity. We found that the brighter the supernova, the slower it decayed. Now we had an empirical way of solving the distance uncertainty. Almost overnight, supernovae became a precision instrument for measuring distances to remote galaxies.

This increase in precision led to the realization that the most distant supernovae were systematically dimmer than nearby ones. The accepted interpretation is that the universe has been accelerating. Hubble's diagram—in particular, Hubble's law that velocity is proportional to distance—must be modified. It applies locally in the nearby universe, but as we peer farther away, a systematic increase in velocity is seen relative to what Hubble's law predicts. At a very large distance from us the supernovae appear dimmer than they would be if the local Hubble law applied. This is acceleration. Because of the acceleration, the distant supernovae are farther away and appear dimmer.

Two Dark Energy Puzzles

There are two immense puzzles about dark energy. Why is it just becoming important at the present epoch of the universe? A not unrelated, but even greater enigma, is this: why is dark energy so small an energy density compared to expectations from the theory of the beginning of the universe?

There is no explanation for the low value of the dark energy density. Any particle physics explanation prefers a value higher by 120 factors of 10. This yields the energy density at the Planck epoch, or the beginning of the universe in terms of classical physics. This energy is considered to be the most natural scale for dark energy. The discrepancy has led to a class of theories in which dark energy increases in density toward earlier epochs, depending on some power of the radiation density. In this kind of theoretical approach, dark energy is no longer the cosmological constant; at least, it is no longer constant. However, there is no compelling reason for this interpretation.

Physicists love to generalize. One can introduce a new parameter, the ratio of the pressure to density of the dark energy. This parameter is conventionally called w. If Einstein's cosmological constant is the explanation for dark energy, w would be –1. However, acceleration only requires that $w < -1/3$. Much observational effort is going into determining the actual value of w and any possible redshift dependence. So far, all results point to Einstein's cosmological constant as being the dark energy.

The Hunt for Dark Energy

This hunt for dark energy will stretch our minds and generate ideas about giant new telescopes on earth and in space. Ongoing and proposed experiments over the next decade are focusing on measuring dark energy by using several independent probes. No single technique may be good enough because there are always systematic errors. Some are known, but the most alarming are unknown. For different techniques, there are different sources of systematic error. Only by combining different techniques can one beat down on the systematics and improve the accuracy of the dark energy measurement. A political utterance by a former U.S. defense secretary, Donald Rumsfeld, in 2002, suggests the kind of exotic challenges involved:

> As we know, there are known knowns; there are things we know we know. We also know there are known unknowns; that is to say we know there are some things we do not know. But there are also unknown unknowns—the ones we don't know we don't know.

Fortunately, the measurement of dark energy is turning out to be like our understanding of musical sound. Primary among the new generation of experiments will be those seeking improvement in the measurements of baryon oscillations. These are the modern counterpart of the music of the spheres. The origin of the density fluctuations, predecessors of the galaxies, occurred in a coordinated way according to inflation theory. Any given scale formed at the same instant. Each scale had its own characteristic instant of creation as the spectrum of density fluctuations developed. Larger scales formed slightly later and were progressively smaller, which eventually triggered the bottom-up formation of cosmic structures. But the synchrony on each scale means that there is a regularity to be uncovered in the pattern of galaxy distribution.

Random density fluctuations in air produce acoustic noise. Harmonization, such as obtained by playing a violin, produces music. As one studies the fluctuations in the distribution of galaxies, a pattern should emerge as the scale is increased. The fluctuations are not random. There is a natural frequency or wavelength in the cosmos that corresponds to the time when density fluctuations first began to grow stronger. The universe was for the first time becoming dominated by ordinary matter. The corresponding wavelength corresponds to the size of the horizon at this moment, or the distance light had traveled since the big bang, which turns out to be about one hundred megaparsecs—the fundamental scale. There are also higher harmonics. Weaker fluctuations are expected at half the scale.

This may seem far-fetched, but we have measured acoustic fluctuations in the cosmic microwave background. These are viewed at some three hundred thousand years after the big bang. The matter fluctuations must show a similar pattern. By viewing density fluctuations at a redshift of 1 in the galaxy distribution, we are probing the epoch of the universe when dark energy is beginning to be a dominant force. The natural wavelength is stretched by the effects of acceleration. Baryon oscillations provide a unique way of measuring acceleration and hence dark energy.

Uncovering this pattern in the galaxy distribution involves

surveys of thousands of square degrees to obtain redshifts of millions of galaxies. Redshifts are needed because they are used to extract distances by Hubble's law of expansion. In this way, one has a three-dimensional reconstruction of the universe. The volume surveyed extends to a redshift of 1 or beyond. This involves much telescope time, and the data acquisition challenge can be partly circumvented, at a price, if spectra are not obtained. A series of filters suffices to give a redshift at lower accuracy than is obtained spectroscopically. The ultimate experiment requires spectroscopy, essential for precise redshifts. Very large telescopes are needed in this case, in order to extract spectra from the faint images of remote galaxies.

Supernova hunting provides a kind of probe that is complementary to measuring redshifts. The light curves of a certain type of supernovae are universal. With enough supernovae, and millions will be needed, one can obtain more and more precise distance measurements. The idea is that with many galaxies at a redshift of 1 or even beyond, one can infer precise average distances at a given redshift. Even now, it is the difference between this luminosity-based distance and that inferred from Hubble's law that indicates that the universe is accelerating. But just how much is it accelerating? Acceleration is how one can measure dark energy, which need not necessarily be just Einstein's cosmological constant. If dark energy is dynamic, the universe might be accelerating less or more than the more naive assumption would predict.

Astronomers have an embarrassment of riches when it comes to possible experiments that search for dark matter. We'll look at three of these. One promising method involves directly mapping dark matter. The basic principle is that gravity bends light. This is precisely the technique first employed to verify Einstein's theory of general relativity in 1919. Light rays from distant stars, passing near the edge of the sun and observed during a total eclipse, were found to be bent by the sun's gravity. The displacement measured amounted to a few seconds of arc.

The same principle applies to light from remote galaxies passing through an intervening cluster of galaxies. The cluster is thought to be dominated by dark matter. There is enough dark matter to distort the light and modify the shapes of distant galaxies. The cluster acts like a transparent gravitational lens. The galaxy images are slightly elongated. The effect is small, but measuring thousands of galaxy images enables one to construct a map of the dark matter distribution. This has to be combined with another technique in order to get a handle on dark energy.

A second approach is complementary and involves X-ray studies of galaxy clusters. One can measure the baryon content of clusters. In this way, the ratio of baryons to dark matter can be inferred. However, we already know what this ratio must be. The light element abundances were produced in the first minutes of the big bang. A successful explanation fixes the baryon abundance. One can, by combining the dark matter and baryon studies, infer the dark matter content of the universe. The result is that a third of the total mass-energy of the universe is in dark matter. The rest must be dark energy, or at least in the form of the cosmological constant. The cosmological constant is constant dark energy.

A third approach measures something slightly different. Surveys are underway that count the numbers of clusters at different redshifts. Theory predicts that more and more massive clusters grow with time. Dark energy reduces the growth rate by countering gravity. There should be fewer clusters. Hence, cluster number counts are a probe of dark energy.

Galaxy clusters can be counted by four independent techniques. One can perform deep galaxy surveys at optical wavelengths. These look for the overdensity of galaxies in a cluster. There are surveys of galaxy shapes that search for the slight distortions produced by gravitational lensing, the curvature of light rays as they pass through dark matter concentrations. These surveys are designed to measure galaxy image distortions for thousands or even millions of galaxies. One has to average more than hundreds of galaxy images to detect the tiny, 1 percent, image distortion produced by an intervening cloud of dark matter. A cluster of galaxies is seen as a region where the distortions are especially concentrated. X-ray surveys look for diffuse sources of X-rays emitted by the hot gas in a cluster. Finally there are microwave surveys. These study the cosmic microwave background as seen through a cluster. The microwave photons scatter off the hot gas and gain energy. So the cluster leaves an apparent hole in the microwave background.

All of these techniques have systematic limitations. There are the known unknowns and certainly the unknown unknowns. However, the systematics differ for the different techniques. We will progress by advancing simultaneously on all fronts, and thereby beating the systematic errors. Nevertheless, using current telescopes, only a modest improvement in limits on the uncertainties in the amount of dark energy is expected in the next decade. A significant reduction in the uncertainties requires a major (and expensive) effort. New types of dedicated telescopes are needed. The projected costs are astronomical.

One project involves building a spectrograph to simultaneously measure thousands of galaxy spectra using an eight-meter-diameter telescope. The field of view on the Japanese Subaru eight-meter-diameter telescope, intended to be used for this purpose, extends over a degree on the sky, or about twice the size of the full moon.

This kind of optical telescope costs a modest amount compared to designs on a giant scale, such as a kilometer-diameter radio telescope. The construction of a telescope this large involves building an array of hundreds of smaller dishes. Spread out, this collective ensemble would have a collecting area equivalent to a square kilometer. Such an array is designed to simultaneously map thousands of square degrees in the sky. The discovery potential is thousands of times larger than any current telescope. In the near future, we look forward to the full operation of the Square Kilometer Array, which will be located in a country in the lower southern hemisphere. Its aim will be to obtain spectroscopic redshifts in the twenty-one-centimeter line of atomic hydrogen for nearly every gas-rich galaxy in the observable universe. By 2020, it is expected to have searched most of the sky to a redshift of about 1, when the universe was half its present size. Its projected cost is at least $1 billion.

However, even these costs can be dwarfed by the potential costs of space telescopes. It is essential to go into space for measurement of infrared and X-rays from distant galaxies. The ultraviolet is also only attainable from space. One example of a successful space telescope that provided ultraviolet, optical, and near infrared data on distant galaxies is the Hubble Space Telescope. One of its biggest advantages over ground-based telescopes is the precision of its images, which have up to ten times higher angular resolution than is currently obtainable from the ground for faint galaxies. The Hubble Space Telescope has cost at least $5 billion. When initially contracted, it was for a price of half a billion dollars. This sort of price inflation is common in space projects because of unexpected obstacles that often arise. In the case of Hubble, these included an optical defect in the original telescope that rendered out-of-focus images. This had to be corrected, and various cameras had to be replaced. Each repair mission requires a space shuttle flight and a dedicated team of trained astronauts.

For pursuing dark energy, a large, new, space-based survey telescope will eventually be needed that is dedicated to one or more of the dark energy techniques we have described. These include hunting for supernovae, mapping dark matter with weak lensing, and evaluating the baryon wiggles.

How Constant Is the Cosmological Constant?

Ever since Einstein developed his ideas about the universe, the notion of a cosmological constant has been famous and much disputed. We still confront two major issues with the cosmological constant—its small value, which corresponds to a very weak force, and the fact that this force has only recently asserted itself in the history of the universe.

First, why is the cosmological constant so small? A natural value of the vacuum density was attained when the universe last went through a phase transition. In fact the quantum gravity theory tells us that there is more than one vacuum, but only one is truly empty, the true vacuum. As it cooled down from the quantum gravity regime, the universe spontaneously split into false vacuua. These false vacuua all have high energy fluctuations in terms of variations about zero. Our universe contains a true vacuum. The true vacuum has infinitesimal fluctuations that actually have been measured.

But a false vacuum has other problems, and it is not a pleasant environment to be in. The false vacuum is a universe containing equal amounts of matter and antimatter—not a universe we would wish to inhabit. The true vacuum resembles our universe, predominantly made of matter. The transition between different states of matter occurred much as when a frozen lake melts and patches of ice remain for a while. The energy density predicted for the true vacuum, the one of our universe, is estimated from the energy at the time of the phase transition. It is higher by about 120 factors of ten than the vacuum we inhabit. This is a major paradox.

The small value is only the start of our puzzlement. Second, we find that the effect of the cosmological constant came late in cosmic history. The acceleration caused by the energy of the vacuum has only recently asserted itself. If it had happened much earlier, galaxies would never have formed. Why now?

Physicists have no answer to these two profound questions: the small value of the cosmological constant and the epoch when acceleration began. Perhaps it is the luck of the draw. In other situations, we might not be around to make any observations. Consider the orbit of the earth. Were it much closer to the sun, we would fry. Were it much further, we would freeze. There is a narrow zone of habitability. We have no theory of why earth is in this zone: it just is. If it were not, we would not exist.

Is the choice of the value of the cosmological constant any different from the coincidence of the earth ending up as being habitable? Actually, it is. In the case of the earth, we have somehow been selected out from many other planets to be habitable, thanks to the particular conditions found on the earth. Some three hundred extraterrestrial planets have been discovered. Almost all are hostile environments for life. So we know that there is a selection principle at work that stems from the physics of planet formation and the biology of the origin of life.

The cosmological constant is very different: we are dealing with only one universe, not many. If we appeal to cosmic selection, then it follows that vast numbers of alternative universes must exist with different values of the cosmological constant. This could be the case. Many physicists are convinced that one can deduce the existence of a meta-universe from this line of reasoning. Others say it is hogwash, or more politely, a step too far. They prefer to wait for a new theory of physics that will provide a fundamental explanation for the observed value of the cosmological constant. This could well be the case. After all, physics can explain the mass of a star and even a galaxy in terms of fundamental constants. What it cannot yet do is explain the values of the fundamental constants themselves. In other words, how fundamental a constant is the cosmological constant? Is it even constant? Indeed, could it have changed over time?

Astronomers are mounting experiments to look for possible changes in the cosmological constant, as well as in the fundamental constants of nature. The results so far are not encouraging for any possible variation over time in the cosmological constant. All measurements point to Einstein's value, both today and in the past.

However, there are hints that the fundamental constants of nature may have varied by small amounts. The data are controversial, as not all observers agree. But the essential argument is precise and well posed. The argument is based on certain atomic transitions in the spectra of quasars at high redshift. The wavelengths of some of these transitions depend on the fine structure constant. Others involve the ratio of electron to proton mass. By measuring these spectral lines and comparing them with results from laboratory spectroscopy, one can search for possible variations with time. There are terrestrial experiments that use natural radioactivity of long-lived isotopes to probe possible variations. The best known of these is an experiment that occurred in nature long ago. At a site in Africa that is naturally rich in uranium, an explosion occurred about a billion years ago.

In 1972, a natural nuclear reactor was found in Western Africa in the Republic of Gabon by a French mining geologist while assaying samples for the Oklo Uranium mine.

The area was naturally loaded with uranium fuel by river sedimentation. The fissionable isotope Uranium 235 was found to be depleted by amounts that could only be explained if it had been used up as fuel in a fission reaction about a billion years ago, when 235U, which has been decaying ever since, made up a much larger percentage of natural uranium than occurs today. Modern power reactors are artificially enriched to a similar level. Abnormally high amounts of fission isotopes were also found in nearby areas that can only be produced in thermonuclear explosions. By studying these rare isotopes, physicists are able to constrain possible variations in the fine structure constant over billons of years. To do better, one has to go into space by using distant galaxies or quasars. The lapsed time for light from remote objects in the universe is enormous: some 10 billion years. Over this timescale, one can be confident that these constants are the same to within one-hundredth of a percent. But there are indications of possible variations at just below this level of sensitivity. If confirmed this would be remarkable news. It would provide possible evidence for the existence of extra dimensions. The interface with our four dimensions would be subject to an infinitesimal leakage of information that could manifest itself as just such a tiny variation in the fundamental constants with time.

Verification of the effect will require higher precision data taken with future generations of telescopes. The best candidate is the Extremely Large Telescope, proposed to have a diameter of up to forty-two meters in a design study by the European Southern Observatory. A planned spectrograph for this telescope will not only be able to search for possible variations in the fundamental constants at a precision of a few parts in a million, but it will be able to directly measure the expansion of the universe. This requires measurement of shifts of relative velocities between distant quasars of better than one part in a million per year.

CHAPTER 9
Eminent Missteps in Cosmology

COSMOLOGISTS are no different from other scientists. They develop obsessions based on intuition that sometimes leads them to strike gold. Persistence is crucial. Albert Einstein's life is a case study of how persistence against conventional views led him to success in his theory of gravitation. It equally demonstrates the weakness of an obsessive approach in that there is no turning back. He failed to unify gravity because he lacked adequate tools.

Einstein was neither the first nor the last to demonstrate such doggedness. In 1600, Giordano Bruno, fifty-two years old, was burned in the Campo de' Fiori in the center of Rome. He is best known to posterity for insisting in his belief that vast numbers of other earthlike planets existed. He was right. Fred Hoyle believed to the end of his life in his steady state theory. He was wrong. Persistence does not always lead in the right direction. But without persistence, we would be much the poorer. That is the story we find in the lives and pursuits of three of our most colorful cosmologists, Albert Einstein, Arthur Eddington, and Fred Hoyle. They represent some of the great missteps in cosmology.

EINSTEIN'S FUTILE QUEST FOR A GENERAL THEORY

Albert Einstein had a vision that seems unique in human history. It involved a mixture of science and imagination. For him, "The true sign of intelligence is not knowledge but imagination."

One of his principles was simplicity: "Any intelligent fool can make things bigger and more complex. . . . It takes a touch of genius—and a lot of courage to move in the opposite direction." But he was careful to set limits: "Everything should be made as simple as possible, but not simpler."

He was inspired by the beauty and truth of mathematics—"Politics is for the present, but an equation is for eternity"—but had a wonderful gift for explaining his theories: "It should be possible to explain the laws of physics to a barmaid."

Born in 1879 in Wurttemberg, Germany, Albert Einstein was to become the greatest physicist of the twentieth century. He graduated from ETH, the Swiss Federal Institute of Technology Zürich, but was unable to find an academic post. In 1902, he became a technical assistant examiner at the Swiss Patent Office in Bern. His job did require a certain knowledge of physics. It was to assess the practicality of patent applications. Not only did he come to grips with the physics of often poorly described new inventions, but he even on occasion corrected their errors in design. He did well. Within two years, he had a permanent position. Despite, and perhaps because of, this career which certainly did not tap his full intellectual powers, he found the time in 1905 to write several articles that had the effect of revolutionizing modern physics.

Hindsight suggests that at least three of these papers deserved separate Nobel Prizes. During his later Berlin period, he received Nobel recognition for one of these, a paper interpreting the photoelectric effect, which is the interaction of light with certain materials that leads to generation of a small electrical current. Curiously, Einstein never accepted the quantum mechanical implications of his photoelectric discovery. His other breakthrough work was on Brownian motion (the random motions of molecules that are buffeted by quantum mechanical forces exerted by neighboring molecules) and on special relativity, which connects measurements in time and space in different inertial frames and, as a corollary, equates matter and energy ($E=mc^2$).

He became world famous eventually for his work on relativity. But in each of the areas that he pioneered, Einstein's approach was unconventional. He was a hard-core theoretical physicist. He began with experimental results that had baffled scientists for decades or even centuries. He managed to explain them by importing and extrapolating ideas from theoretical physics to their logical consequences.

In 1914, Einstein became director of the Kaiser Wilhelm Institute for Physics in Berlin. Over the following year, he gave a series of lectures at the Prussian Academy of Sciences on his new theory of gravitation. In this theory, Newton's law of gravity was replaced by what Einstein called general relativity, which described the large-scale universe as a curved fabric of space-time. Special relativity asserted that the laws of physics did not depend on the frame in which an observer is moving. This is kinematics. The bold new step was to generalize this approach to include frames that were accelerating. This included the effects of gravity, as would be the case on the earth or near the sun. For this to work, space no longer was Euclidean. Gravity was equivalent to having a curved space. Light no longer travels along straight lines. Acceleration is represented by curved light paths, or geodesics. The way we feel gravity is by acceleration.

In this model, the curvature of space-time replaces gravity, a radical change. Yet Newton's law of gravity remains central to the new theory as long as gravity is not too strong. Indeed, as Einstein himself put the case, somewhat poetically, "To the Master's honor all must turn, each in its track, without a sound, forever tracing Newton's ground."

Cosmology soon sprang out of the new theory. As noted earlier, it was pioneered by the contributions from Friedmann and Lemaître, and even adopted by Einstein once the observational support for Hubble's law was established. Einstein was at first a reluctant convert, as was Hubble, to the new cosmology. It did not take long, however, thanks largely to the persuasive advocacy of

Eddington and Lemaître, before the expanding universe theory gained general acceptance.

Einstein himself soon moved on to other matters. He had not forgotten relativity theory, of course. As he once quipped, "When you are courting a nice girl an hour seems like a second. When you sit on a red-hot cinder a second seems like an hour. That's relativity."

But his heart was elsewhere. Einstein spent the last forty years of his life on a new quest. His aim was to unify the forces of gravity and of electromagnetism. The battleground was quantum mechanics, a theory that Einstein never fully accepted, despite being one of the founding fathers thanks to his work on the photoelectric effect.

In the 1920s, the original quantum theory of Niels Bohr with its mechanistic view of the atom was replaced with the probabilistic theory of quantum mechanics. Einstein would not accept that probability described reality. He retained his conviction that physics had to describe the laws that govern reality. After all, he had succeeded with atoms, photons, and gravity. He was unwilling to abandon faith in his ideas. In 1926, Einstein wrote to Max Born:

> Quantum mechanics is certainly imposing. But an inner voice tells me it is not yet the real thing. The theory says a lot, but does not really bring us any closer to the secret of the Old One. I, at any rate, am convinced that he does not throw dice.

Einstein sought new insights into the nature of gravity. It had to somehow meld with electromagnetism at high enough energy. He tried to combine a universal law of gravitation with the electromagnetic force. He sought a unified mathematical approach to the two fields that encapsulated gravity and electromagnetism. He searched for what we would call a unified field theory. Unification is supposed to describe all forces as different manifestations of a single underlying force. (Recall our example of the Rolls-Royce and Volkswagen being melted by the big bang heat into unified chemical components.)

Einstein looked for an explanation that would be deterministic and not probabilistic. This surely was a fundamental error of judgment. As with his previous successes, he was now guided, and indeed obsessed, by theoretical prejudice. But this time, it was not seasoned with longstanding experimental anomalies like those he had previously been able to brilliantly visualize as hints of, and incorporate into, a new theory.

Einstein simply would not accept the Copenhagen interpretation of the quantum theory held by the Danish physicist Niels Bohr of Copenhagen and his younger German associate, Werner Heisenberg. Their quantum theory invoked a probabilistic, nonvisualizable account of physical behavior. Einstein was not persuaded. Rather than delve into the philosophical aspects of the theory, which was to inspire whole generations of philosopher-scientists, he sought instead to find gedanken (or thought) experiments that might show the quantum theory to be internally contradictory. His attempts were doomed to failure. Not until fifteen years after his death were the strong and weak nuclear forces unified, both in theory and experiment. However, the unification of the small forces with the large-scale force of gravity still remains an elusive goal. Yet Einstein's dream survives. It drives the current quest for the ultimate theory of everything. One manifestation of this quest is string theory. This is a theory of quantum gravity that portrays the properties of fundamental particles and forces as stringlike shapes in a higher dimensional universe.

As a consequence, Einstein became increasingly isolated in his research toward generalizing the theory of gravitation. This work ultimately turned out for him to be a wasted effort. For more than half his scientific career, he is widely considered to have been marginalized. He became the Don Quixote of physics in his fruitless search for a unified theory of electromagnetism and general relativity. His big mistake, with hindsight, was to denigrate and even ignore the role of the quantum theory of particle interactions.

Einstein was a prominent figure in politics and society. He was frequently called upon to give opinions on a wide variety of controversial issues. Despite the subsequent publicity, he remained a modest figure throughout his career. He fully realized his limitations, saying that "the man of science is a poor philosopher."

Einstein's relation to God is one of agnosticism, bordering on belief in some intangible higher authority. He was certainly not an atheist. He did not hesitate to advertise his belief that religion was generally a good thing, although he never really defined what this meant. Typical was his famous statement, "Science without religion is lame, religion without science is blind."

He admired the key religious figures: "What humanity owes to personalities like Buddha, Moses, and Jesus ranks for me higher than all the achievements of the inquiring and constructive mind." His God was a social and benevolent God, but one that remained remote from individuals. "I cannot accept any concept of God based on the fear of life or the fear of death or blind faith," Einstein said. "I cannot prove to you that there is no personal God, but if I were to speak of him I would be a liar." Nevertheless, it is clear that science led Einstein to his God

> Everyone who is seriously involved in the pursuit of science becomes convinced that a spirit is manifest in the laws of the Universe—a spirit vastly superior to that of man. . . . In this way the pursuit of science leads to a religious feeling of a special sort, which is indeed quite different from the religiosity of someone more naive.

What, then, about morality and belief? This did not really enter the God equation for Einstein. His God was not necessarily a measure of morality. As he said, "My religiosity consists in a humble admiration of the infinitely superior spirit that reveals itself in the little that we, with our weak and transitory understanding, can comprehend of reality. Morality is of the highest importance—but for us, not for God."

This left humanity to blame for all adverse consequences. We are responsible for any immoral applications of science.

Eddington's Ghost of a Fundamental Theory

Sir Arthur Eddington was one of England's greatest theoretical astronomers of the twentieth century. He made immense contributions to our understanding of the inner structure and evolution of stars. He pioneered our understanding of the stability of stars. He realized that if a star weighed in at more than a hundred times the mass of the sun, its pressure would be so dominated by radiation that the star would blow itself apart. But in addition to his understanding of stars, he closely followed Einstein's work and was one of the first to appreciate the significance of Einstein's new theory of gravitation.

Eddington was born in 1882 in Weston-super-Mare, Somerset, in the United Kingdom. He was a precocious mathematician. Not only did he complete the Mathematics course at Cambridge in two years instead of its normal duration of three years, but he was top of his class. He went on to become director of the Observatories at Cambridge, where he led astronomy research from 1914 to 1944.

He was one of the first to appreciate the significance of Einstein's theory of general relativity, which he first encountered in 1916. The new theory provided an explanation for a well-known astronomical phenomenon that was not explained by Newton's theory of gravity. This was the advance of Mercury's perihelion, the point at which it orbits closest to the sun. Einstein's explanation predicted the bending of light rays passing close to the sun. This could only be tested during a total eclipse of the sun. Eddington fortuitously had an opportunity to undertake a key eclipse measurement. He came from a Quaker background and was a conscientious objector in the First World War. The resulting exemption from war service led to Eddington's being appointed in 1919 to lead an eclipse expedition to the island of Principe in West Africa. There, in the visual path of totality of the eclipse, he would be able to verify a key prediction of Einstein's theory.

The idea was to observe the positions of stars on the sky that were close to the sun. Ordinarily they would be invisible against the bright glare from the sun. But in the three-minute duration of a total eclipse, one could see these stars and measure their slightly displaced positions. These could be compared with their normal positions, long after the sun had set. Gravity curves space, and the apparent stellar positions should be correspondingly displaced if the light paths from distant stars passed close to the sun.

As the sun moved around the sky, the star positions would revert to their undistorted positions. Newton's theory of gravity actually predicted a small displacement. This is because light particles, or photons, have energy. But Einstein's theory predicts twice the displacement. In March 1919, Eddington sailed from England for sunnier climes. By mid-May, he had set up his instruments on the island of Principe. But drama intervened. The eclipse was due to occur at two o'clock in the afternoon of May 29. However, that morning there was heavy rain. Eddington wrote in his diary:

> The rain stopped about noon and about 1.30. . . . We began to get a glimpse of the sun. We had to carry out our photographs in faith. I did not see the eclipse, being too busy changing plates, except for one glance to make sure that it had begun and another half-way through to see how much cloud there was. We took sixteen photographs. They are all good of the sun, showing a very remarkable prominence; but the cloud has interfered with the star images. The last few photographs show a few images which I hope will give us what we need.

Eddington did manage to obtain several photographic plates. Over the next week, Eddington developed the plates. Almost all were of poor quality, but he eventually recorded in his notebook, "One plate I measured gave a result agreeing with Einstein."

To posterity, this was an astounding result, given the poor quality of Eddington's data. In the end, he only had two plates of sufficient quality to give a significant result. A similar expedition to Brazil to view the same eclipse had far better observing conditions and generated more than twenty plates. However, more than half of these gave a larger value than predicted by Einstein's theory. Eddington realized that smearing by the atmosphere would tend to exaggerate the displacement he was seeking. This gave him a good reason for dismissing the larger values. So he rejected the outlying points and only used the results from the South American expedition that agreed with his data.

What Eddington could not do was be sure that his own data was unaffected by atmospheric smearing and so would give an artificially high value of the stellar displacement. But when he presented the results to the Royal Society of London in 1919, he did not discuss this possible source of uncertainty. The eclipse observations provided the first confirmation of Einstein's theory that gravity bends space. This made banner headlines in the war-weary world of 1919.

Eddington wrote, in a parody of the *Rubaiyat of Omar Khayyam*:

> Oh leave the Wise our measures to collate
> One thing at least is certain, light has weight
> One thing is certain and the rest debate
> Light rays, when near the Sun, do not go straight.

As a footnote, it is worth adding that more than ten subsequent eclipse expeditions failed to confirm Eddington's result until new technology came to the rescue. Modern data, however, confirm Einstein's theory.

Eddington was a lucky man. He obtained the correct answer from data that with hindsight was vastly inadequate. He was strongly motivated by theory, and strongly felt that observation had to confirm theory, if the theory was correct. Without confirmation, theory was merely metaphysics. He penned a maxim which makes this plain enough: "For the truth of the conclusions of physical science, observation is the supreme Court of Appeal."

In his later years, Eddington strayed from this maxim. Perhaps he knew better, or perhaps his luck did not hold. At any rate, a more fundamental theory beckoned.

He was fascinated by the fundamental constants of nature, and understanding their origin became the new focus of his research. He was not hindered by so-called experts who had pondered similar problems for years. Indeed he seems to have believed that familiarity with the history of a subject was a hindrance to creative research in that area.

Astronomer H. C. Plummer wrote of him:

> A bold imagination was coupled with an exceptional knowledge of those features which are accessible to observation. . . . To launch out into unknown seas, to be venturesome even at the risk of error, Eddington felt himself called, and the reward of the pioneer came to him.

Eddington sought the ultimate theory of physics. There had to be a unifying theory that related the fundamental forces. He attempted to explain various numerical coincidences between the large and the small measures of the universe. Most of his ideas were published posthumously in his monograph *Fundamental Theory* in 1946. His goal was to unite quantum mechanics and general relativity (large-scale gravity). To achieve this, his approach was to determine the relation between the sizes of different physical systems.

His attempts at a fundamental theory amounts to his version of what we now would call a theory of everything. This theory led him to assert:

> I believe there are 15 747 724 136 275 002 577 605 653 961 181 555 468 044 717 914 527 116 709 366 231 425 076 185 631 031 296 protons in the universe and the same number of electrons.

This number is 136 times 2 to the power of 256 (136×2^{256}). Unfortunately, few took his theory seriously. Indeed no one subsequently has even been able to reproduce Eddington's logic and calculations. Not only did Eddington's ideas receive little acceptance, but they were seen by many to verge on scientific mysticism. What went wrong with this great astronomer? For one thing, he was out of his depth in attacking the quantum theory. He may have been inspired by the British mathematical physicists Paul Dirac's success, but lacked the physics weaponry to compete on a similar level.

A sampling from some of Eddington's other biographers gives us a sense of his scientific legacy. We can begin with the account of C. A. Ronan in *Their Majesties' Astronomers*:

> Eddington, hard-headed mathematician and down-to-earth astronomer though he might be, possessed a mystical side to his nature and the last years of his life were spent in an attempt to construct a huge relativistic synthesis of the physical universe, an edifice in which the bricks would be subatomic and astronomical evidence of the observer and the mortar the underlying mathematical relationships between them. But none of these efforts worked out, and in fact his results were irreproducible.

Next we have this from MIT physicist I. H. Hutchinson:

> The practically unanimous verdict of scientific posterity is that Eddington, like Einstein, failed. The major difference between them, though, is that Einstein was well

> aware of his failure, whilst Eddington thought to the end that he was well on the way to succeeding.

Finally, and far more pertinent, Einstein himself offers this epitaph on Eddington: "His imagination was not adequately balanced by his critical facilities." He and Einstein were equally single-minded.

Hoyle in Troubled Waters

Fred Hoyle dominated British astronomy in the latter half of the twentieth century. He was a brilliant theoretical astronomer. Perhaps the brashness of his Yorkshire character and the unorthodox paths he pursued in his later career kept him from getting the ultimate recognition of a Nobel Prize. He was born in 1915, went to Cambridge to study mathematics, and wrote a thesis that he never submitted for a PhD degree. The PhD was a relatively new innovation at Cambridge. It was not universally appreciated in the first half of the twentieth century. Hoyle's talent was eventually recognized when he was appointed as professor of mathematics at the end of the Second World War.

Hoyle made his reputation in studying how stars work. He solved one of the greatest puzzles in astronomy at that time. This was the origin of the chemical elements. He explained how thermonuclear fusion reactions proceeded in the core of a star and led to its eventual death. German American physicist Hans Bethe had just developed the theory of nuclear reactions involving hydrogen burning at very high temperature. He showed that stars were powered by the thermonuclear conversion of hydrogen into helium. Once the star exhausts its core hydrogen, Bethe realized, it begins to contract because the pressure in the core becomes insufficient to support the mass of the star. The stellar core heats up under the inevitable action of gravity, which sets the stage for burning the helium. Hoyle demonstrated that the helium turns into carbon, which itself burns into heavier elements such as oxygen. Even heavier elements are formed, all the way up to iron. But once the iron core forms, no more thermonuclear energy is available. Once there is no longer a source of internal heat to combat gravity, the star must collapse.

The collapse eventually triggers a gigantic explosion, or supernova. At the end of a star's life, the heavy elements are blown out into space by the explosion. The energies attained in some parts of the explosive debris are so high that nuclear reactions can recommence. These are rather special reactions, relying on neutron bombardment of the stellar material. During these explosions, elements heavier than iron can form. We find that essentially all of the chemical elements heavier than helium that are present in the universe were formed in exploding stars.

Hoyle's reputation preceded him. He gave many public broadcasts and wrote popular books. He even wrote outstanding science fiction novels such as *The Black Cloud* and *A for Andromeda*. In his scientific research, he often challenged accepted dogma. He became known as an unconventional astronomer who dared to challenge the establishment in science as well as in science politics. While Hoyle's work on the origin of the chemical elements was widely recognized as a breakthrough, the Nobel Prize, given in part for this work and awarded many years later, went to his collaborator, nuclear astrophysicist William Fowler. Perhaps the Nobel committee's choice reflected a distaste for Hoyle's unorthodox approach to cosmology.

In 1948, Hoyle ventured into cosmology. He initiated the concept of a universe that lived forever: "Things are the way they are because they were the way they were." Hoyle detested the big bang theory. Indeed, in one of his BBC radio lectures, he coined the name as a derogatory gesture. The name not only stuck, but this rival theory was to greatly outlive and supersede Hoyle's steady state vision of the universe.

Hoyle could not accept that the universe had such an extreme origin. The big bang postulates that the universe began in a highly, indeed almost infinitely, dense state that lasted the tiniest fraction of a second. It expanded and cooled down from this almost singular and hot state. Such an origin is required by advocates of the expanding universe theory.

This seemed so far from his intuition that Hoyle preferred to think of the universe as being the same everywhere and at all times. As the galaxies moved apart, new galaxies appeared. To Hoyle and a small group of friends, creation of new particles from nothing—which purported to explain the "expansion" of the universe—seemed a more reasonable hypothesis than extrapolating back to a state of quasi-infinite density. He developed the steady state hypothesis further with the aid of his co-discoverers Hermann Bond and Tommy Gold, and with younger colleagues such as Jayant Narlikar. The steady state universe emerged from all of this intellectual effort. In the 1950s, it was a serious challenger to the big bang cosmology. A side benefit was that the steady state universe not only had no beginning, but it also had no end.

Unfortunately, with time, the steady state universe failed challenge after challenge from the observers. First there was the discovery of radio galaxies. Hoyle's great Cambridge rival, Martin Ryle, was developing the new field of radio astronomy just down the road from Hoyle. Ryle discovered a vast population of radio sources in the distant universe whose number increased markedly with survey depth. This had to be the first solid evidence for the evolution of the universe, according to Ryle. The universe was denser in the past.

Still, Hoyle refused to abandon the steady state universe that allowed no dense past. He responded to Ryle by asserting that we live in an immense and underdense region, a void, in the universe that is characterized by a deficit in the number of galaxies. This brash assertion was harder to counter since at this time most of the radio sources were unidentified. A curious afterthought is that the current hottest topic in cosmology—that the acceleration of the universe provides evidence for dark energy—can perhaps also be countered if we live in a vast void. History goes in circles.

To return to the steady state versus big bang saga, there was a clear turning point. The death knell for all but the diehard steady state supporters came with the discovery of cosmic microwave background radiation in 1964 by Penzias and Wilson. Hoyle tried desperately to conjure up needlelike dust grains in space that would be effective microwave emitters. However, the data proved impossible to explain by anything but a big bang origin.

Nonetheless, the steady state controversy did inspire one of the most important discoveries in astronomy, namely how the chemical elements originated. In the early days, Hoyle's greatest opponent on the issue of the age of the universe was George Gamow. We recall that Gamow had fathered the hot version of the big bang. He was motivated by the need to account for the origin of the chemical elements by thermonuclear reactions. He seized the big bang as the ideal environment where the required high densities of matter would naturally occur. What was perhaps less natural was the billion-degree temperatures required to initiate the fusion process. As we have seen, Gamow brilliantly sidestepped that issue. He regarded it as an entirely plausible initial condition for the universe. Thus, he focused on the prediction that, since the universe must have been hot, there ought to be a relic temperature today, subsequently detected as the relic radiation glow of the cosmic microwave background.

In fact, Gamow made a notable mistake in his logic that Hoyle and others seized upon to their advantage. In his early writings, Gamow believed that all of the chemical elements were made in the big bang. Gamow made the elementary mistake of not recognizing that massive stars were short-lived in calculating their contributions to heavy element production (such as oxygen, carbon, and iron). So his estimated production of heavy elements fell short by a large amount. But in fact, there were many generations of these stars, as first realized by U.S. astrophysicist Edwin Salpeter.

In contrast to Gamow, and correctly so, Hoyle developed the alternative hypothesis that the heavy elements were made in the interiors of massive stars that subsequently exploded as supernovae. With his collaborators, nuclear physicist William Fowler and astronomers Geoffrey and Margaret Burbidge, Hoyle developed the complete theory of the origin of the heavy elements in debris from exploding stars. This theory is accepted to this day. In fact, a central part of Gamow's theory has also found universal acceptance. The light elements were made in the first minutes of the big bang. His predictions of the universality of the light element abundances, and of a relic radiation field, contributed two major pieces of evidence for the big bang.

We know now that both Gamow and Hoyle made lasting contributions to our understanding of the origin of the chemical elements. Gamow's hypothesis accounted for the abundances of the lightest elements, while Hoyle and his collaborators explained the abundances of the heavy elements—significantly, the carbon and oxygen that make life on earth possible. Hoyle's frustrated venture into cosmology led him into an even more radical idea. Once an idea took hold, this prototypical Yorkshireman was not easily diverted. He explains the mentality thus:

> To achieve anything really worthwhile in research, it is necessary to go against the opinions of one's fellows. To do so successfully, not merely becoming a crackpot, requires fine judgment, especially on long-term issues that cannot be settled quickly.

A century earlier, the Swedish scientist Svante Arrhenius had developed the theory of panspermia, which asserted that life exists throughout the universe and that life on earth originated from the propagation of bacteria in comets and asteroids. Hoyle's dust grains turned out to be promising sites for growing organic matter. He interpreted astronomical features seen due to dust absorption against stars as complex material that hosted bacteria and viruses. New viruses arrived from space to cause epidemics of influenza. He studied English public school health records and inferred that some epidemics were so spontaneous and widespread that cosmic intervention was required. Epidemiologists scoffed, but Hoyle always continued to seek more evidence for his theory. As observations of the spectrum of fossil radiation improved, Hoyle's alternative hypothesis became harder and harder to maintain. In the end, his theory was only abandoned after its key proponent himself died.

CHAPTER 10
The Universe in Seven Numbers

THE FIELD OF COSMOLOGY has divided itself into two pathways. One of these is guided strictly by data and description, the phenomenological approach. The other pathway is in search of an ultimate theory about the beginning and the end of the universe. Different scientists decide which route they want to pursue, but invariably these two paths cross each other, as we've seen in many of the chapters of this book.

Cosmology today is a product of both data and theory. At this point, it is worth our while to summarize what we know about our universe based on the data and the theories of the past several decades. Once we have this in hand—and indeed we can simplify what we know into just seven numbers about the universe—we can see how cosmologists speculate about its ultimate beginning and its very future. We first go back to the beginning.

The question of the beginning has ancient roots, as we can discern from the ponderings of the book of Job, composed probably twenty-five hundred years ago. "Where is the way to the dwelling of light?" Job asks. "And as for darkness, where is the place thereof?" Today, our giant telescopes are like time machines that let us travel into the past. And looking back in time means looking far away. How far can we go? Let's start nearby. The sun is nine light minutes away. The nearest star is four light years distance. The center of our galaxy is twenty-four thousand light years away. Andromeda is 2 million light years distance. The Andromeda galaxy is a naked eye object, though we need telescopes to get its distance.

The most distant galaxy discovered to date was found with a combination of the world's largest telescopes on mountaintops in Chile and Hawaii. It is 12 billion light years away. These telescopes have mirrors with diameters of eight or ten meters. They are big and expensive pieces of machinery, but provide the size necessary to collect enough light to see the most distant objects.

By pushing back the frontiers of our exploration of the universe, we can understand our own planet in relationship to the largest scales of reality. We know, for example, that the earth formed 4.6 billion years ago. We use the oldest rocks as fossils to date the earth, and we use meteorites to date the solar system. We can see back even further. Our Milky Way galaxy formed 10 billion years ago, which corresponds to the ages of the oldest stars. The chemical abundances in these stars are primitive compared to those in our sun. These stars are fossils that help us date the Milky Way itself.

We can see back even further. We have discovered that the universe is expanding. Distant galaxies are rushing away from us. We also know that our planet, our sun, and even our Milky Way galaxy are hardly at the center of things. Just imagine any point on the surface of an expanding balloon. This point, any point, is the center of a two-dimensional expanding curved surface. When the earth formed, the universe was about two-thirds of its present size. When our galaxy formed, the universe was about a third of its present size. The light from the most distant galaxy known was emitted when the universe was around a tenth of its present size. Then galaxies were ten times closer together. The night sky was full of galaxies.

But this cannot go on indefinitely. In fact, we see fewer and fewer bright galaxies as we look even deeper. We knew this long before the age of the big telescopes. When observation fails us, we turn to theory. Theory tells us that galaxies cannot be ever present earlier and earlier. Otherwise the night sky would be glowing with galaxy light. This is Olbers' paradox, named after German astronomer Heinrich Wilhelm Olbers in 1823. Olbers' paradox was known much earlier, to Johannes Kepler in 1610, and described by English astronomer Edmond Halley and Swiss astronomer Jean-Philippe Loys de Cheseaux in the eighteenth century.

Even Edgar Allan Poe wrote about Olbers' paradox, and he even described the modern solution:

> Were the succession of stars endless, then the background of the sky would present us a uniform luminosity, like that displayed by the Galaxy—since there could be absolutely no point, in all that background, at which would not exist a star. The only mode, therefore, in which, under such a state of affairs, we could comprehend the voids which our telescopes find in innumerable directions, would be by supposing the distance of the invisible background so immense that no ray from it has yet been able to reach us at all.

We infer that the galaxies, all of them, including those unseen, have a finite age. As we peer back in time, we must enter the dark ages, before there were any stars. But even the dark ages are not completely dark. There is a feeble glow in microwaves that comes from this period. Turn on your television. Put it out of tune, to a channel that is not broadcasting. One percent of the fuzz on the screen is the glow from the big bang. One of the greatest scientific discoveries ever was the fossil radiation from the beginning of the universe.

What else do we know? Thanks to the early astronomers, we know our universe is incredibly dynamic. Edwin Hubble discovered that the universe is expanding. He obtained galaxy distances and galaxy spectra. Redshifts measure velocity. The farther away a galaxy, the greater its speed. Modern data shows that the universe is expanding at a rate of ten miles per second for every million light years of distance. Now we can extrapolate back in time. The universe began 14 billion years ago. That in itself is a major conclusion. But there is more.

Once, in the past, the universe was as dense as the sun. George Gamow's great insight was to predict that it must also have been as hot as the sun in order for thermonuclear reactions to make the light elements. Helium makes up a third of the mass of all the matter in the universe; the helium was mostly produced in the big bang. The feeble residual glow of the big bang was discovered by serendipity the year before Gamow died, and when his theory was long forgotten as well. That is science—the interaction of theory and data, even when theory is sidelined for a while.

Gamow's theory was confirmed by Arno Penzias and Robert Wilson, who were unaware of his ideas. They were radio astronomers mapping the Milky Way with a discarded telescope that was one of the first used for satellite communications. They discovered an excess microwave glow emanating from the whole sky. It was not of local origin (bird droppings was one hypothesis that they quickly discarded), or terrestrial, or solar, or galactic. By a process of careful elimination they found that the glow came from the universe. This was the fossil relic from the once-fiery big bang. The universe around us was once a thousandth of its present size. This was the point in time at which the relic radiation came into visibility, three hundred thousand years after the beginning of the expansion. The universe became transparent. Before then, the universe was shrouded in a dense fog of radiation.

Once, the universe was so dense and hot that even particles could not exist. This was a mere nanosecond after the big bang. Our current theory of particles is being tested by a consortium of scientists at CERN using the Large Hadron Collider. The gigantic machine might allow us to study the state of the universe as it was a trillionth of a second after the big bang. But cosmology does not stop there. When our machines reach a limit, we use theory.

According to quantum theory, as used by cosmology, the universe began at a tiny fraction of a second. We can be precise: the universe began at a thousandth of a trillionth of a trillionth of a trillionth of a second after the big bang. Before then, we have no adequate theory, although we speculate in a later chapter on the ultimate beginning. At the incredibly high energies involved, the fundamental forces of nature are no longer distinct. The nuclear forces cannot be distinguished from the electromagnetic force. Only as the universe expands and cools to a lower energy state do these forces separate.

There are two consequences of this cooling and separation of forces. One is that protons should even today be ever so slightly unstable. The process is very slow at today's low energies. But it must occur, if very rarely. Look at enough protons and one will decay before our eyes. Suppose a proton were to decay after 10^{20} years. This is around 3 times 10^{27} seconds. This would be the average lifetime of a proton. What this means is that some protons are decaying somewhere every second. Now, look at our own bodies. How would proton decay affect them? Our body contains something like 10^{29} protons. There would be something like thirty radioactive decays per second in our body. The proton decays into gamma rays among other things. With so much irradiation on our body tissues, we would rapidly contract cancer. We infer that protons last at least 10 billion times the age of the universe. Otherwise we all would be dropping dead of cancer.

Proton decay is inevitable if the fundamental forces once were united. The weak and strong nuclear forces would then have been equal, and quarks freely formed and decayed. One can test the validity of the principle of grand unification by detecting proton decay today.

An amazing experiment was developed to measure proton lifetimes. Take a vat of water containing at least a million trillion trillion molecules, or about ten tons of water. Proton decay is statistical. It can happen any time. On the average it takes a very long time. But in ten tons of water, there would be a decay rate of around one proton per year if the proton lifetime were a million trillion trillion years. A proton decay is an energetic event, producing gamma rays. These can be looked for. The current limit with experiments that monitor fifty thousand tons of purified water is about a thousand times longer. The first such prediction from the simplest model of proton decay has been tested and eliminated. It predicted that protons decay in a million trillion trillion years. If even ten decays per year were measured, this would point to a proton lifetime that is a hundred times longer, or around 10^{32} years, 10^{33} being the number of protons monitored in the experiment. This is where we might expect to find evidence if nature is sufficiently accommodating. The experiments are continuing.

Another consequence of the application of grand unification theory is that we can understand how, when the universe cooled, it underwent a transition in phase from symmetry to asymmetry. This transition releases vast amounts of latent energy, much as when ice melts to water. The result is that the universe expanded dramatically, driven by this release of energy. This is the inflation we saw in chapter 3. For a brief moment, the universe inflated enormously.

We can now understand why the universe is so large. We also can understand why it is so uniform, relatively speaking. It looks the same everywhere. To imagine this, think of how a balloon gets smoother when it expands. And as a bonus, we even can understand the origin of the tiny fluctuations in density that seeded galaxy formation. They are relics of the quantum era. These earliest shifts in density are seen as temperature fluctuations in fossil radiation. We use the measurements of these fluctuations to run computer simulations to observe the effects of gravity as the universe expands. From an initially almost homogeneous distribution of density, structure is seen to develop. Regions with a slight excess in density accrete surrounding matter. The large-scale structure of the universe develops. Galaxies and galaxy clusters form.

When it comes to telling this story of how the universe evolved, there is one slight complication: we don't yet know what dark matter is. What we can do is measure its existence. For example, our computer simulations are reliable for the development of the dark matter distribution. This is because only gravity is involved. Dark matter constitutes 90 percent of the matter distribution. But again, what is the actual dark matter particle made of? We think it probably consists of weakly interacting particles that can self-gravitate according to Newton-Einstein gravity. Astronomers are beginning to map out the dark matter distribution by the technique of gravitational lensing. The dark matter clumps act like transparent lenses that bend light and distort the images of background galaxies. The simulated and observed maps of dark matter are in excellent agreement. So our assumption about what dark matter is cannot be too far from reality.

Just Seven Numbers

The universe seems highly complex. Yet overall its cosmology is remarkably simple. It takes just seven numbers to describe a model that fits all modern data. This includes all we know about the cosmic microwave background and the distribution of galaxies.

Our first number is the age of the universe. This is determined to an accuracy of a few percent once we measure the rate of expansion.

Our next four numbers are the densities of four important components of the universe. These are the densities of ordinary baryonic matter, of radiation (or photons), of dark matter, and of dark energy. Overall, we can directly map the densities of radiation and ordinary matter with telescope surveys that produce all-sky maps of galaxy distributions and the microwave background. In these sky maps, we measure the densities and the spatial variations from place to place. By reconstructing the surveys from the data in two dimensions, we can measure to the three dimensions we inhabit. What we don't see in this galaxy survey are the baryons that make up the intergalactic gas, the dominant home to baryons. But we can measure these, too: we study the X-ray glow of the intergalactic plasma. We have two techniques for measuring dark components that dominate the density. One uses precise distance determinations, measuring the geometry of the universe (or, equivalently, the combined density of its various components). The other looks at the rate of growth of density fluctuations. Because dark energy is unchanging and dark matter is reduced in density as the universe expands, we can infer their densities.

Our last two numbers are the values that controlled the seed density fluctuations from which all structure emerged. We arrive at this knowledge of structure by first taking the strength of the density fluctuations that are responsible for forming the galaxies. To this value we add another important number: the spectral index that describes the distribution of the density fluctuations.

That's an even seven! Armed with these seven numbers we go from the generality of Einstein's theory of gravitation to a model that gives a near perfect fit to everything we know about the universe. Of course, assumptions are made en route. The most notable of these is that the universe is homogeneous and isotropic (that is, having both a smooth consistency and also seeming to have the same structure in all directions). But it is a great triumph of the theory that we can fit such diverse data with so simple a model. After all, there are thousands of independent data points that are needed to represent the matter and radiation distributions we observe, yet seven numbers suffice.

Forecasting the Future

From this model, with its seven numbers, we can predict the probable future of the universe. It is likely that its acceleration will continue forever. That at least is what the model tells us. In other words, this is what we can deduce from the data. There could be new data some day that might change our model. Or there could be eventualities that are not foreseen in the simplicity of the model.

Here is one possibility. The dark energy could spontaneously decay. In this case, the acceleration would be halted. But there is not a shred of observational evidence that leads us to this conjecture.

Instead, we must face the forecast of continued acceleration of space, which means that within our horizon, or how far we can see with the largest telescope, there will be fewer and fewer galaxies. Today, the world's largest telescopes can map some 10 billion galaxies. But there will come a day, a trillion years or so in the future, when our horizon will span a single galaxy, our own. A lonely future beckons. Of course by that time, the only stars remaining will be the dimmest of dwarfs. The initial hydrogen required to fuel fresh star formation will have been completely exhausted.

However, there still will be the occasional flicker from stars self-destructing in supernova explosions. This occurs because pairs of white dwarfs can take immensely long times to spiral together before merging into the explosive mixture of a Type I supernova.

This is a trillion years in the future. After this, there are no longer any stars capable of shining by nuclear energy. There is a feeble glow from the many white dwarfs which are becoming blacker and blacker as the final vestiges of gravitational energy are released. The universe is dark. It is full of stellar relics and smaller lumps of planetary mass. It will stay this way for a very long time—in fact, for about a number of years that equals 10 to the power of 33, or a billion trillion trillion years. This is our best guess of the lifetime of a proton. The unification of the fundamental forces of nature requires that protons must be exposed to the radioactive decays that characterize the weak nuclear force. They are highly suppressed inside a proton, but if we wait long enough, decay is inevitable. That at least is what the simplest viable unification models of particles predict. At some point in time, possibly within a billion trillion trillion years, the protons will decay. All that will be left in the universe are electrons, neutrinos, and photons. There will no longer be any structures. Nothing could exist that remotely resembled a planet. Life in any imaginable form would be over.

Is the Universe Infinite?

How can we tell whether the universe is finite or not? This has been a topic of great speculation. For example, nineteenth-century English naturalist Thomas H. Huxley wrote, "The known is finite, the unknown is infinite; intellectually we stand on an islet in the midst of an illimitable ocean of inexplicability. Our business in every generation is to reclaim a little more land."

When it comes to an infinite universe, our attempt to claim a bit more territory poses interesting issues. Anything could happen, and did happen. The writer T. H. White, applying his Arthurian theme, says simply that what is not limited may indeed happen: "Everything not forbidden is compulsory," he writes. These ground rules provide a happy hunting ground in an infinite universe.

First of all, what is infinite? Bigger than you can imagine! But we can never prove the universe is infinite. If it were truly infinite, there are some very strange contradictions that arise. There is another London or New York out there on a distant planet with identical inhabitants doing identical things. Except that the weather is better. At its worst, there are infinite numbers of London cities. We may never access such wonders as may exist if they are beyond our horizon, present or future. It is more scientifically plausible to ask whether we could ever prove the converse, namely that the universe is finite.

On the grounds that an infinite universe is philosophically unappealing and quite possibly rather boring, let us compromise and consider that the universe is just very, very large. This is what we believe and indeed, more to the point, what we measure. Hence, we have a nearly infinite universe. So how do we arrive at this conclusion?

Cosmologists could find a signature that would point to a finite universe. There is an imprint of finiteness in the cosmic microwave background radiation. It disappears if the universe is too big, but it could someday provide a measure of finiteness. If the part of the universe that we observe is not special in any way, then the pattern of fluctuations we see in the radiation on the sky should extend to scales as large as the universe. This includes what we cannot see.

You may wonder how we can measure what we cannot see, and

of course we cannot. But we can measure the bits of such trajectories that lie within our horizon. Imagine coming across the steps to a place called heaven. Of course you can't see this heaven, or even get to heaven without dying first, but the steps are there. So, one could measure accessible pieces of fluctuations that are in their totality invisible. And the absence of pieces, in a finite universe, would leave a distinguishing signature in the microwave sky. We haven't found such evidence yet, needless to say, although there are curious hints in the data that such an effect of finite size could exist. Better experiments are needed to study the tiny fluctuations in the temperature of the microwave sky.

CHAPTER 11

Our Place in the Universe

Cosmology leads inevitably into considerations that have philosophical and even theological overtones. Inescapably, cosmological theory is confronted with the question of why human beings exist in this particular universe, and what is our role as observers? After all, if we did not exist, who cares? The physical models of the universe have enough chinks in their armor for these philosophical, and even metaphysical, questions to squeeze through the cracks.

For centuries, the traditional answer to human existence has been that we are more than an accident in the universe. In recent decades, this idea generally has been called the anthropic principle, described in the opening chapter. Such ideas are very persuasive for theologians, of course. The anthropic approach has also proved fascinating, if not always convincing, to philosophers. For cosmologists, however, this can often be troubling territory to enter.

In science, we try to stick to data. We try to explain the physical origins of things according to laws and principles. Indeed, we hope we can explain by physical law and theory what otherwise seem to be amazing "accidents" or coincidences. One of these is the rise of human life on a planet some 14 billion years after the big bang. Galaxies, stars, and planets did not necessarily have to exist, and yet they are here—making our existence possible.

The scientific project of explaining our presence has been around quite a while. Over the centuries we have worked our way from a geocentric universe, beloved by the Greeks, to a heliocentric universe, pioneered by Nicholas Copernicus. For obvious reasons, we have called the idea of our noncentrality in the universe the Copernican principle. Up until Einstein's era, the Milky Way was taken to be the entire universe. This conclusion was overthrown by Hubble's pioneering measurement of the distance scale of the universe. It became clear that our Milky Way was a run-of-the-mill galaxy. Neither large nor small, but depressingly average. In relationship to the entire universe, the nonexceptional status of our environment has been called the principle of mediocrity. Our changing view of the Milky Way galaxy illustrates how we arrived at this knowledge of our averageness.

Average or not, cosmologists still grapple with our existence. To settle this issue, we need some criterion for normality. Otherwise, we could propose any kind of bizarre universe or universes and fall over the edge—into science fiction. The easiest way to begin is by being fairly down to earth, asking about our physical existence in relationship to the laws of physics.

Our Physical Size

The smallest person in the world, an Indian named Gull Mohammed, was 22 inches in height. The tallest, an American named Robert Wallow, had a vertical span of 8 feet 11 inches. The average height is 5 feet 4.5 inches, a respectable height for much of the world's population. But the observed distribution of human heights fills only a small part of this range. Why is this? And where are the giants, as in the mythological world of Jack and the Beanstalk? And where are the Lilliputians?

We know the answers. Gravity makes life most unhealthy above a certain height. Our bodies could not be supported and we would topple over. And dwarfs prefer to mate with other dwarfs. In view of the brutal competition for a foothold on earth, they too are presumably destined for extinction below a certain body and brain size. Hence, genetic evolution with an assist from gravity eventually narrows the height range to the observed distribution. We do not need to invoke either the anthropic principle or the intervention of a Great Designer to reconcile observation with theory. Biology and physics provide the answers. But remember, we are only talking about our existence on the earth, a relatively small speck in the cosmos. We cannot rely on gravity and biology to explain our existence in relationship to the entire universe.

When cosmologists rely on the anthropic principle to explain why we exist, most use a weak version, while a minority use a strong version. The weak version is used by cosmologist Alex Vilenkin of Tufts University. He says simply that the universe is the way it is because we are here to observe it. This is a tautology, a circular argument, but it still has some merits for some kinds of explanations. This approach assumes the existence of the so-called multiverse, in which many possible universes exist that are inhospitable to our existence. The latest theories of quantum gravity count some 10^{500} realizations of possible universes, all of them different from each other because their various fundamental constants of nature differ.

On the one hand, given the staggering array of alternative universes in the multiverse, it becomes exceedingly improbable that our observed universe should even exist. Nevertheless, it does exist, and according to the weak anthropic principle, our existence has simply selected the right universe. After all, we can only observe a universe of a certain size, old enough for stars and planets and life to have developed. But is this kind of idea physics, or is it metaphysics? Princeton cosmologist Robert Dicke went so far as to say that, because of our existence, the age of the universe must be old enough to allow stars to synthesize carbon, the basis of biological life. The next logical step is to point to how the values of all of the fundamental constants of nature, which may vary throughout the multiverse, are determined by the requirement of our presence.

Another approach to the weak anthropic principle, preferred by many of my colleagues, selects only the small subset of so-called pocket universes within the multiverse, and this pocket allows galaxies to form and life to develop. Within these finite pocket universes, the probability then becomes large for scientists finding only a small, but nonzero, value for dark energy today.

A strong version of the anthropic principle claims that intelligent life is inevitable somewhere in the multiverse. However, these strong anthropic arguments are undercut by the inclusion of a possibly infinite age for the universe (which suggests that the universe expands and contracts forever and we are merely seeing one of the expansions). Much can happen over a long time in a universe that perpetually renews itself via eternal inflation.

In all of the anthropic arguments, proponents attempt to explain the relationship of human existence to the values of the fundamental constants of nature. There are at least three rival hypotheses on how these values came to be. The first said the values could have been selected by a Grand Designer. This has great appeal to proponents of the intelligent design of the universe. Most cosmologists argue forcefully that there is no need to invoke such a concept, although ultimately it reduces the problem to a question of personal belief.

The second option appeals to currently unknown physics. For example, we saw that the height distribution of human beings can be understood by known rules. Perhaps we simply do not yet know the rules for navigating in the multiverse. It may be that it takes an infinite time to populate the plenitude of landscapes. Rolling down the hill of Alex Vilenkin's false vacuum, we have no reason to talk about probabilities of our universe to appear. The universe began in a symmetrical state. It was the inflation afterward that generated the highly asymmetric state of matter around us now. The implication is that it would take so long for a vacuum to produce this particular universe that it would happen only once. The question of probability is irrelevant.

The third alternative, and to my mind the most likely resolution, says that there was no selection at all. We are here because we are here. This is what must happen in an infinite multiverse. Some versions of quantum gravity appeal to the complexity of the initial conditions to assert that there is an infinity of landscapes and universes in the multiverse. If this were the case, the game—and the mystery—is over. The dice were rolled and our universe was inevitable, somewhere in the multiverse. And here we are.

Are we unique? That is another story, although an infinite multiverse universe would allow an infinite number of universes like ours, just as an infinite universe allows infinite numbers of stars, planets and earths.

Remarkably, one can test this hypothesis. Future experiments will measure the curvature of space with exquisite precision. If the curvature turns out to deviate from flatness (that is, an ever-expanding universe), we would come to a conclusion unprecedented in human thought. A slightly closed universe would prove the finiteness of space. A slightly open universe would go far toward demonstrating that space is infinite, at least in the most generic of cosmologies. Were this to turn out to be the case, one would no longer need to invoke any anthropic principle. It is simply redundant.

Darwinian Selection or Eternal Inflation?

Some cosmologists simply say that, from the start, the anthropic principle is not even scientific. That is because there is no way to test how the universe might differ if we did not exist, or to compare our universe with others. This is the view of American quantum cosmologist Lee Smolin, for example. As an alternative, he pioneered an idea of cosmic evolution that is similar to the Darwinian principle in biology. Given the vast multiplicity of universes, Smolin says, perhaps there is a process of natural selection by which one universe becomes preferred. To explain the evolutionary principle, Smolin appeals to black hole formation as a means of spawning new universes.

This process of new universes popping into existence is not for-

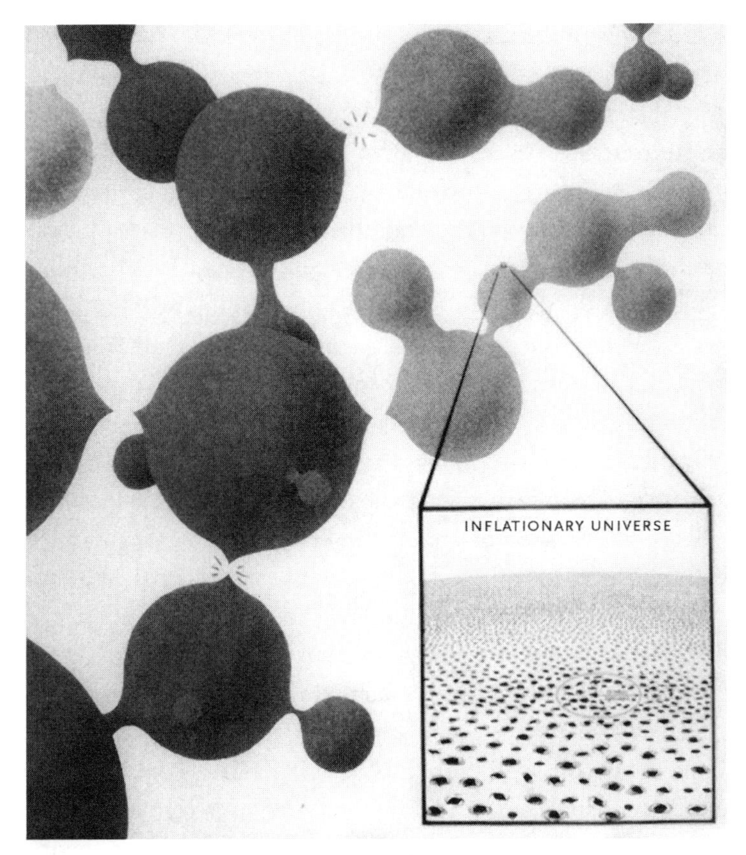

Figure 11.1. The birth of baby universes. (From E. Mallove, "The Self-Producing Universe" *Sky and Telescope* [September 1988].)

bidden by physics, so therefore we are allowed to theorize that it may occur. If it does occur, prolific black hole formation presents a possible channel for universe creation. In this proposal, the winner in this cosmic race—that is, the most probable universe—is the one teeming with black holes. Our universe is not far from this state. We know that at least 0.01 percent of baryonic matter is in the form of massive black holes. These per se are not ideal for universe creation, as they normally are very long-lived. It is their precursors that might hold the secrets of creation. It is even conjectured that the elusive dark matter is in the form of black holes, in which one has a universe whose black hole content provides more than 20 percent of its mass-energy content. The massive black holes could be the observed byproduct of the final state of a black hole–dominated universe.

Black hole birth may provide a physical process of cosmic natural selection. However, the underlying hypothesis that transient black holes might spawn universes is highly speculative. For example, very small black holes are believed to evaporate. The smaller the black hole, the greater the curvature of space in its vicinity. If a black hole weighs less than the mass of a mountain, something like 100 billion tons, the fabric of space near its surface is ripped apart and the black hole spews out mass. It evaporates. Such tiny black holes are called mini-black holes. Short-lived mini-black holes dominated the universe at the Planck instant. Singularities are at the centers of all black holes. These are potential gateways to other universes. But it seems unlikely that evaporation of mini–black holes would result in the generation of relic universes. Rather all of the information once contained in the black hole is evaporated along with its mass.

A chief rival theory to this "birth by black holes" scenario is the idea of eternal inflation of the universe. Inflation can replicate itself. It can happen again and again. That is because quantum theory tells us that the most improbable events can occur if we wait long enough. Similarly, if the universe is large enough, such improbable events must occur somewhere. So inflation can set in anywhere at any time in a truly large universe. This is eternal inflation. It means not only that we are not in a special place, but neither is our observed universe a special universe. It is the ultimate successor to the Copernican principle.

You may wonder whether there are any consequences of eternal inflation. One of the greatest paradoxes in cosmology is that the universe is dominated by something called dark energy. All we know is that the universe is accelerating, and this mysterious energy force is responsible. (Recall, it was first postulated to exist by Einstein to stop the universe from collapsing, before the expansion was discovered.) However, the value of dark energy is remarkably small. The quantum gravity theory applied to cosmology predicts that the existence of this small of a force actually is highly improbable. One solution is that there are a vast number of universes produced by theories like eternal inflation, with all possible values of dark energy. Ours just happens to be the one in which galaxies and stars could form. If dark energy were large, stars could not form. Gravity is unimportant compared to repulsion. The other universes were too hot or too cold, or too dense or too rarified.

Is There a Multiverse?

Theoretical physics grapples with the idea of a multiverse. The ensemble of an infinitude of universes including our own local realization is a bizarre concept. Is it even in the realm of physics to postulate the existence of a multitude of coexisting but noncommunicating universes?

There are several theories for the origin of a multiverse. Still, it is a sad reflection on our current state of knowledge that none of them are any more convincing than that attributed to a frequently cited old lady who stood up at the back of a lecture on the nature of the universe and angrily denounced the eminent lecturer, American philosopher William James. She asserted that our Milky Way galaxy rests on the back of a giant turtle. This is in fact an ancient Hindu myth. And what pray is the turtle standing on, queried the bemused lecturer? She replied that the answer was obvious: "There are turtles all the way down."

Now there are several choices. One is that the universe, our universe, is unique. It is just that our minds have not yet grasped the ultimate theory for explaining this. Another possibility is that there is a multiverse, within which are found all manner of coexisting but noncommunicating universes. There is an infinitude of choices. But almost all of these are inimical to the evolution of life:

too hot, too cold, too lumpy, too smooth, too young, too old, and so on—hence, what Arizona State University physicist and author Paul Davies has dubbed the Goldilocks enigma. Just one universe is selected by our presence as observers. Such multiverse theories are alleged to be, writes philosopher Neil Manson the last resort for the desperate atheist. More practically, the underlying physics is incomplete: we have absolutely no idea if the many so-called landscapes in the multiverse can be populated by pocket universes, as its denizens are dubbed.

The existence of a bio-friendly universe is a vastly improbable event that does seem to demand an explanation. One option is to appeal to higher authority, or intelligent design, to explain our presence. This opens up the question of who designed the Designer, and why. Nor does it distinguish monotheism from Valhalla, the realm of the gods in Norse mythology. While this solution is a guaranteed conversation stopper for science-minded people, the choices do not stop here. Perhaps the universe is a vast Gaia-like entity (named for Gaia, the Greek goddess of nature) with built-in awareness of its purpose. Or maybe we are living in a *Matrix*-like computer simulation. One might infer that Hollywood came closer to the truth than the most eminent string theorists (whose string theory tries to explain quantum gravity). This in itself need not surprise anybody.

There is one solution that I find most appealing. Suppose the universe were infinite. All bets are off in this case because it is logically impossible to compute the odds for or against life in an infinite universe. The most unlikely events occur somewhere. And here we are, in our own bio-friendly pocket universe. Perhaps what is most important for the physicist is that this seemingly metaphysical solution of an infinite universe can be experimentally testable, at least in principle. Such tests primarily involve studying the very large angular scale properties of the cosmic microwave background and also improving the precision of measurements of the curvature of the universe.

The Frontier Limits of the Multiverse

We know the earth is not flat, because no one has reported falling over the edge. The roundness of the earth limits its size, giving it a frontier that cannot be crossed. We can take a similar line to the multiverse, an approach that purports to explain many of the so-called coincidences in nature. Beyond certain boundaries in nature, drastic changes take place. Many of these changes could be a catastrophe for human existence, blocking it from taking place, and those are the changes in which this kind of approach is interested. The laws of probability push us toward the largest space imaginable. For example, we need to understand why the proton mass is only very slightly larger than the neutron mass. This mass difference coincides with the scale of the theory of electro-weak interactions. If it were even slightly different, stars would not form. So we are likely to be pressing against this particular frontier. The proton-neutron difference is close to what is observed if ordinary stars exist.

Let's pursue further the laws of probability. These tell us what is possible and what is not. Cosmologists who find a hard boundary that separates different realms of values for the fundamental constants may go on to argue that similar logic might separate universes. At least one of these contains galaxies. In the usual view, galaxies form from small fluctuations that grow stronger by gravitational growth. These are suppressed if the cosmological constant is too large. Galaxies would not form. It is an all-or-nothing view. Observers populate galaxies. The pending catastrophe of no stars, no galaxies, confronted with the pressure from the multiverse, isolates us in a corner of the multiverse where conditions are just right. The advantage of the catastrophe approach is that it sets a hard boundary in superspace. The new wrinkle in this view, expounded by Lawrence Hall of the University of California at Berkeley, is that the combination of multiverse pressure and catastrophic frontiers provides an explanation of many of the fine tunings found in nature.

Are We Typical or Are We a Cosmic Fluke?

Here is another way around the Goldilocks dilemma. Perhaps we are far from typical as observers of the universe. It has been argued that quantum theory cannot require us to be anywhere, such as here, with high probability. The Copernican principle, that the earth is not the center of the universe, returns with a vengeance. Even our very existence could be an unlikely event. American physicists James Hartle and Mark Srednicki calculate that

> [it] is perfectly possible (and not necessarily unlikely) for us to live in a universe in which we are not typical. . . . Cosmological models that predict that at least one instance of our data exists (with probability one) somewhere in spacetime are indistinguishable no matter how many other exact copies of these data exist.

This leads to remarkable implications. Experiments in cosmology might become pointless because there is no logical correspondence between us and the way the universe looks. Canadian cosmologist Don Page concludes that "observations would count for nothing in distinguishing between theories, and much of cosmology would cease to be an observational science."

Now, there is another interpretation of the Hartle-Srednicki conjecture. We would no longer be able to use odds (that is, probabilities) to distinguish between universes. What difference would it make if our universe were one in 10^{500} if we could no longer argue whether this was likely or unlikely? It had to happen somewhere, and it might as well be us. This type of reasoning puts an end to naïve applications of the anthropic principle, which purports to explain our special place in the universe.

The anthropic observer's role can be questioned more deeply as well. Perhaps there are virtually no observers. It may be that all but a handful of the 10^{500} vacuums that populate the landscape are simply unpopulated by observers. Physics is irrelevant, since the quantum theory requires an observation for reality to be manifest. Pity Schrödinger's cat, named for the thought experiment of the German physicist Erwin Schrödinger. In this experiment, the cat is in a box and forever condemned to be both alive and dead in a dual virtual existence until the act of observation (by opening the box) is performed. This leads to one of the most well-known paradoxes in quantum theory. Release of poison gas was triggered in the cat's box by a radioactive decay that is inherently probabilistic. The decay occurs, or it may not. But once the cat's box is opened there can be no uncertainty. It is either alive or dead. Physicists scramble to explain this by postulating that there are two cats, one alive, one dead. Only one of these, unpredictably, materializes when the box is opened.

We get into this quandary because of the observer. Take him or her away, and the state of the cat no longer matters. The same could be true for almost all of our unhealthy universes. They are barren of observers. In physics terms, the rolling field which turns the vacuum state into a universe rolls so slowly that it barely arrives at its destination. Only the longest-lived universes get populated. One of these is ours, comfortably old enough to make stars. If only one universe, ours, made it to this desired state, we would be left with far fewer choices to explain our existence. There would be no older universes. This way we avoid what has been dubbed the youngness paradox. Why are we so young when infinitely many universes are far older?

Of course, there still might be other alternatives, such as universes with different values of the fundamental constants. These universes would not even make any stars. But at least there are fewer of them. Our recourse here may be to say that an improved theory of physics will provide values of the physical constants. We are in good company. After all, Eddington thought so. He calculated the fine structure constant to many significant figures from first principles. But he frustratingly obtained the wrong answer.

Confronting Our Greatest Challenge

Dark energy prevents cosmologists from sleeping at night. Why is its density so small, when all particle physics conjectures lead us to a value larger by 120 factors of 10? And why is it just beginning to dominate the expansion of the universe, as the acceleration is taking over? Physicists hate the idea of fine tuning, their expression for adopting what seems an incredibly unlikely event. Hence, they have come up with a proliferation of multiverse speculations. These reek of metaphysics and are impossible to verify, short of science fiction–like advances in time-travel technology. They are probably even impossible to falsify, with any foreseeable experiments.

There is one argument on the strength of the force of dark energy that merits serious attention, however. It purports to explain why the dark energy must be small and recent in its effects. When we measure density inhomogeneities in the universe, we map them into the temperature fluctuations seen in the microwave background. The level of the fluctuations is infinitesimal. It amounts to a thousandth of a percent. If Einstein's cosmological constant, the simplest manifestation of dark energy, were much larger than the measured value, galaxies could not have formed from these fluctuations. However, we live in a galaxy, and we would not be here were there no galaxies. Hence there is an observational bias in our calculations. Our presence requires a low value and a recent domination of dark energy.

Of course, one might ask why there should be any dark energy at all. Galaxies are perfectly happy in a universe without dark energy. Here the case for a cosmological constant, and thus for dark energy, comes from string theory. (String theory is a theory of quantum gravity. It tries to explain how the smallest and largest forces in the universe, and indeed all fundamental interactions, once were united in multidimensional strings before the big bang.) There are many possible values for the strength of the dark energy force. The most probable value is expected to be where the number of model options is largest. This is the greatest value. But this pressure toward bigness is limited by the need for us to observe dark energy and more specifically to live in a galaxy. Hence dark energy is inevitable, small, and only recently dominant. This argument for the smallness of the dark energy was first given by Nobel laureate physicist Steven Weinberg.

There are at least two fundamental flaws in this argument. One comes from statistics. The arguments always follow the precepts of the Reverend Thomas Bayes (namesake of Bayes' theorem). This eighteenth-century statistician inspired a modern form of computing the odds of an event happening, taking into account all possible theoretical possibilities. The larger the volume of possibilities, the more likely it is that one of these options is favored. So we need to count observers. It turns out that if one allows not just for the actual number of observers, but also for the future number in any given universe, the odds shift dramatically. Infinity always wins. When applied, Bayes' method guarantees that near-zero dark energy universes are favored. There are far more galaxies per unit volume combining all the universes with zero and very small dark energy than in those with high dark energy. So the upper bound on dark energy survives, but there is no longer any predictability. There should be no dark energy.

Our attempts to calculate predictable amounts of dark energy only get worse. Even the upper bound on dark energy is not a robust argument. Here is why. If fluctuations could be very large initially, this would weaken the constraint on dark energy. The usual arguments says that if they are large, black holes rather than galaxies would form very early in the universe. However, this argument assumes that the primordial fluctuations are generated by inflation using something called the inflaton field. This is an energy field with no preferred direction. We call such an energy field a scalar field, as opposed to vector fields of energy which are directed. This field provides the energy that causes inflation and generates the quantum fluctuations that seeded galaxies and are seen in the microwave sky. The fluctuations are equivalent to enhancements of matter and radiation. If too large, black holes would form and completely dominate the universe. This is certainly not the case.

Any reasoning about inflation has loopholes. One is that there is no unique energy field that drives inflation. In fact, string theory dictates that there would be a whole slew of scalar fields. The outcome is that the most general primordial fluctuations consist of two different types. One corresponds to compression of matter and radiation. This amounts to a slight increase in the local curvature of space. Einstein tells us that this excess gravity is what causes the collapse eventually of galaxies, once the universe has cooled down and is dominated by ordinary matter. These are the fluctuations expected if inflation is driven by a single scalar field.

But reality need not be so simple. With more than one field responsible for inflation, there is now another type of fluctuation, in which an overdensity of matter is exactly compensated by an underdensity in radiation. There need no longer be a pure curvature fluctuation at all. There are even fluctuations with zero curvature. The name "curvaton" has been given to the combination of fields that can create these zero-curvature fluctuations.

Moreover, because the radiation density is so enormous early on, one can balance this with a huge matter fluctuation. Very large fluctuations in the matter are possible. But because of the balance with radiation pressure, any risk of premature black hole formation is avoided.

On the scales of galaxies, the fluctuations could even be larger by some twenty factors of ten than the usual curvature fluctuations and still lead to acceptable microwave background temperature ripples. Galaxies would form early, as soon as the universe consisted of atomic hydrogen and the radiation stopped scattering against the electrons. This was a million years after the big bang, but the galaxy mass clouds would still evolve and make stars. The resulting universe would not be an exact replica of what we observe, but this hardly matters. The curved universe would contain a multitude of galaxies, stars, and planets. It could have a far larger dark energy content than our observed universe. It would inevitably contain observers. We cannot use galaxy formation to set any significant constraint on dark energy.

A Mathematician's Paradise?

Mathematics may have the last say in selecting our universe. Perhaps certain laws of mathematics that operate in the deepest realms of reality have dictated that we have the universe we have. Such a proof would produce an unassailable result. The standard model of particle physics is a fundamental element of string theory, which, as noted earlier, is a theory of quantum gravity. Particles are described by one-dimensional vibrations in a higher-dimensional field. Strings exist in a ten-dimensional hyperspace. This provides a mathematical description of all of particle physics. Particle masses are intrinsic to the theory, as is supersymmetry. The theory is still incomplete, but it provides the most compelling description of the beginning of the universe.

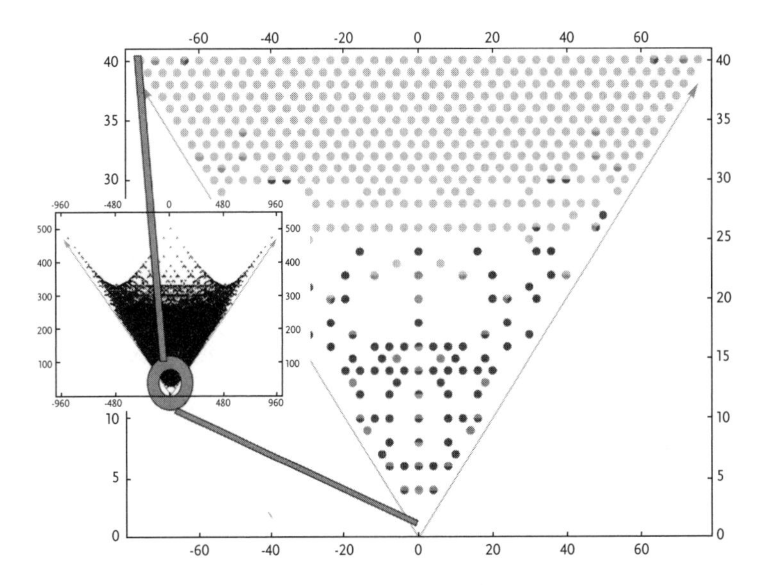

Figure 11.2. How the search for geometrical simplicity can reduce the number of manifolds that are possible in separate universes from 10 raised to the power of 500 down to only 3. (From P. Candelos and R. Davies, arXiv 0809.4781.)

Armed with string theory, one finds a definite number of possible geometric spaces, any one of which could eventually compactify into our universe. If we use supersymmetry as our guiding principle, there are 10^{500} of these spaces, called Calabi-Yau manifolds. Each is a precursor state of a possible universe. Each allows three generations of particles. Hence all of known particle physics can be incorporated into every one of these manifolds. The price one pays is that each has six dimensions. These existed at the Planck instant, at the beginning of the universe, when gravity was united with the other fundamental forces. This was the epoch of quantum gravity.

Our universe has four dimensions, three in space and one in time. As the universe expanded away from the Planck density and cooled, the gravitational interaction became much weaker than the nuclear forces. At the same time, the extra dimensions compactified. One of the manifolds became our universe. There may still be traces on very small scales of hidden dimensions. These would reveal themselves as tiny deviations from Newton's laws of gravity on very small scales. Physicists are searching for such effects but have not found any. There may also be traces of hidden dimensions on very large scales. Not all memory of the pre–big bang universe need have been erased by compactification. Again, no significant deviations from Einstein-Newton gravity have so far been found.

How do we classify the plethora of early universes? The mathematical conjecture is that there must be a sort of topological unity to a viable model. One example is called the Euler number. Think of this as the number of holes in a higher dimensional space. Because of its handle, a coffee mug has an Euler number of one. The simplest space has an Euler number of zero. It has no holes.

We would like to believe our universe is this simple. But if a universe is chosen at random from all of the possibilities, this outcome would be pretty unlikely given the vast number of possible spaces. Most of the predicted Calabi-Yau spaces have very large Euler numbers. The Euler number is a measure of complexity. Oxford mathematician Philip Candelas has proposed an argument that can reduce the number of possible spaces to only three if one selects those with the smallest topological numbers. This is a promising start for limiting the overwhelming number of possibilities confronted at the beginning of space and time, suggesting that simplicity may count for something. It amounts to a geometrical way of obtaining a huge reduction in our possible options.

CHAPTER 12
Cosmology's Future

Is DARK ENERGY worth pursuing as a primary goal in cosmology? This question in turn is motivated by another one: will we ever have a theory of dark energy? On a more practical level, given that we inhabit a galaxy, will there ever be a fundamental theory of structure formation?

While one can always find inflationary models to explain whatever phenomenon is represented by the flavor of the month, the generic predictions, associated with the vast majority of the models of inflation, have certainly had two immense successes.

The Two Successes

One success is the verification of the flatness of space. Another stems from an achievement of the WMAP (Wilkinson Microwave Anisotropy Probe) satellite. The results from the space satellite have almost succeeded in eliminating one of the rival hypotheses to inflation. This prediction, independently due to Edward Harrison and Yaakov Zel'dovich, is of the nature of primordial density fluctuations, which asserts that the fluctuations should have exactly the same strength on all scales. Cosmologists measure the shape of the fluctuation power spectrum. It can be characterized by a single number, the spectral index n, which is predicted to be exactly unity on the basis of simple but compelling scaling arguments.

However, this is one situation where it seems that simplicity has to be abandoned when confronted with reality. The result from the

WMAP satellite is that the spectral index *n* is approximately 0.97, and differs from 1. A value of 1 is excluded. Such a number slightly below unity corresponds to a small change in strength with scale of the fluctuations, with the smaller scales being slightly stronger. This consequence of the finite duration of inflation is expected. Smaller and smaller fluctuations exit the horizon later and later as inflation peters out and the fluctuation distribution gradually rolls over in power.

Nowadays, cosmology seems rather unexciting. All measurements converge on the standard cosmological model with its hypothesized ingredients of dark matter and dark energy, which are themselves poorly understood. Future experiments concentrate on reducing the current error bars, with the possibility always lurking of finding hints of new physics. Confidence that we have found the final solution requires immense hubris, given our woefully inadequate mastery of the first instants of the big bang. The ultimate theory of cosmology will surely include our standard cosmological model as a component.

Time Travel

Could we ever prove any of these conjectures? The multiverse hypothesis posits a vast number of universes. All exist. Only by measurement can we hope to test their existence. We need to go there. Our best bet is to construct a time machine to travel into other universes. This is far out of view in terms of what we are currently capable of doing.

Time travel between universes is a truly remote goal. To get there, we first would need to master time travel within our observable universe, let's say on the earth. No fundamental law of physics forbids time travel. Success is surely a matter of technology. Unfortunately we are very far from design, let alone construction, of a prototype time machine. But let's use our imagination, inspired by the theory of gravitation.

Time travel is possible in principle via wormhole technology. A wormhole exists, according to Einstein's theory of general relativity. It is a bridge into another part of space and time. It can access other universes. Of course, we have never detected a wormhole, but this need not stop our speculations. After all, most of the universe is unexplored.

Time travel is the only way to test the multiverse hypothesis. We need to go there. This may be, and probably is, the realm of science fiction, but it is not excluded by the laws of physics. Traveling far in space necessitates travel in time. Space-time constitutes the fabric of the universe. We cannot separate the two. Time travel can be described in the context of physics. It is a necessary precursor to traveling across the universe. However, travel in time involves apparent contradictions that we need to confront.

It would not be a good idea to travel back in time and kill one's great-grandmother for some alleged sin. There is a contradiction, since we cannot change the present. One response is that invoking quantum uncertainty allows us to avoid this paradox. One can demonstrate this by thought experiments on the quantum scale. Let us imagine we are able to burrow through space by constructing a tunnel, analogous to a train tunnel penetrating a mountain. The material of a space-time tunnel would need to be incredibly strong to prevent it from collapsing under the pressure of the surrounding space with its matter and energy fields. But such a material is no different in strength from what would be required for a spaceship to enter within the horizon of a black hole, a journey that may be required to find the nearest wormhole.

Shooting particles into such a tunnel through space-time would result in their emergence at an intrinsically uncertain point in space-time. The odds of encountering themselves on the way back would be infinitesimally small, so they could not be knocked off course on the way into the tunnel. Whether there is a macroscopic correspondence is not known but seems likely. One would never find one's great-grandmother. She would be hidden in a fog of quantum uncertainty.

Take time travel as a desirable goal for many of us. We all regret past errors. What would we not give to have another chance, to revisit that brief encounter with a person never seen since, or to retake that crucial examination? Or for those who are bored by modern tourism opportunities, why not a photographic safari for Tyrannosaurus Rex in situ, or a walking tour of the original Seven Wonders of the World? Physics says all of this and more is possible. Not today, nor tomorrow, but such adventures lie within the explorable bounds of physics.

Here is the recipe for the tunnel: construct a wormhole. This bizarre object, predicted to exist in Einstein's theory of gravitation, is a tunnel through space and time. It is only accessed through a black hole, so no return is possible without violating the laws of physics. Nothing can escape from a black hole. Quantum gravity, the ultimate theory that unites gravity and quantum theory, and which is the Holy Grail of particle physics today, actually requires that tiny black holes and wormholes must exist throughout space. However, their lifetime is fleeting, and they disappear before we can capture or even detect one. The vacuum is thought to be teeming with such brief-lived oddities. Wormholes are there to be trapped, if only we knew how. The trapping, however, can be done without violating physical laws. It merely requires developing a very stiff material that resists gravity.

Virtual wormholes should exist for fleeting instants, according to quantum gravity theory. Capturing one would require unimaginably strong forces. Entering one may be dangerous: it could close upon itself. Our best bet may be to find a rapidly spinning, massive black hole. Many hundreds of massive black holes are known, often lurking at the centers of galaxies. Even our Milky Way has one. Current astrophysical theory strongly suggests that all black holes are spinning, because most massive stars are spinning rapidly. These stars are the ones destined to form black holes. A few will grow to reach monster dimensions. The one at the center of our galaxy weighs 4 million solar masses.

Only if the final stages of collapse involve rapidly spinning matter can one understand how to find a wormhole. A rapidly spinning black hole has a deformed event horizon. Normally the event horizon prevents the casual observer from accessing the central singularity. An event horizon is impenetrable. For the outside observer, time grinds to a halt. But with spin, the singularity becomes accessible. It can even be naked, with no protective event horizon. One can enter one of these and locate the central singularity where a wormhole might be lying in wait for the unsuspecting traveler.

The technology for a spaceship to survive passage into the event horizon of a black hole requires material of unbelievable strength. As one approaches the event horizon, the tidal forces become immensely strong. Any known material would be ripped to shreds. But sufficiently stiff material should exist, at least in principle, to survive the voyage. Indeed, the matter in a neutron star has the required strength to resist being pulled apart by the gravity near a wormhole. Fashioning a spaceship out of such material borders on fantasy today, but science fiction has a habit of becoming reality over the course of a few decades. Witness the writings of Jules Verne and H. G. Wells.

Even stranger material that counters gravity is found in highly diluted form. We call this the dark energy that is responsible for the observed acceleration of the universe. Time machine technology would require us to harness this force, use it to prop open a wormhole before it closes up, and widen up the wormhole to a large enough opening for the intrepid time traveler to jump in and vanish from visibility. She would end up, if she survived the journey, in a new region of space and time, either in our universe, or—far more likely, given the quasi-infinitude of possible destinations—in some other universe. Multiverse travel has become feasible. Of course, getting back to the place from where one set out would set a further challenge to our intrepid voyager. But perhaps she is only too willing to escape to new horizons, especially if there is little choice in the matter.

At least one technical problem must be overcome first, however. Theory suggests that once a wormhole opened, it would instantaneously accrete matter and close in on itself, which represents a serious challenge for our future wormhole engineers. A better strategy might be to find a preexisting wormhole at the center of a rapidly spinning black hole.

Technical issues aside, this type of time travel is our best bet for exploration of space-time. Unfortunately, it is far afield from any technology we can presently imagine—classified by the American academic and impossibility expert Michio Kaku as an advanced technology of Type II. Such technologies are achievable in perhaps a million years without violating any of our cherished concepts.

Possible Surprises

When we think about the future, the words of the ancient natural philosophers offer some perspective. Roman philosopher Lucius Annaeus Seneca (d. c. AD 65), for example, wrote in his *Natural Questions*, book 7:

> The time will come when diligent research over long periods will bring to light things which now lie hidden. A single lifetime, even though entirely devoted to the sky, would not be enough for the investigation of so vast a subject. . . . And so this knowledge will be unfolded only through long successive ages. There will come a time when our descendants will be amazed that we did not know things that are so plain to them. . . . Many discoveries are reserved for ages still to come, when memory of us will have been effaced. Our universe is a sorry little affair unless it has in it something for every age to investigate. . . . Nature does not reveal her mysteries once and for all.

Humanity has possessed advanced technology as we now think of it for less than a century. Progress has been staggering. We have entered the realm of what the ancient Greeks ascribed to the gods.

We have mastered flying machines and space travel, achieved almost instantaneous video communication at vast distances, and controlled explosions of virtually unlimited power. Imagine what we might be capable of achieving in a century, in a thousand years, or in a million years. It would be a rash person who stated that our knowledge of physics precluded our being able to master a time machine. Of course, there always have been established scientists—as liable to error as the rest of us—who express their certitude in areas they would do well to avoid. Consider the following examples:

> Heavier-than-air flying machines are impossible.
> —Lord Kelvin, president, Royal Society, 1895

> If [quantum theory] is correct, it signifies the end of physics as a science. —Albert Einstein

> Anyone who expects a source of power from the transformation of these atoms is talking moonshine.
> —Physicist Lord Rutherford, 1935

> There is not the slightest indication that nuclear energy will ever be obtainable. —Albert Einstein

Nor were physicists alone in having such pontification skills. Others joined in, such as the astronomers:

> Space travel is utter bilge! —Richard Wooley, 1956

> Space travel is bunk. —Harold Spencer Jones, 1956

These quotes from the leading lights of the British astronomical establishment (Jones was Astronomer Royal, a post created by King Charles II in 1675, from 1933 to 1955 and Wooley from 1956 to

1971) came one year before the launch of the *Sputnik* satellite on October 4, 1957. To paraphrase physicist Lev Landau, "Cosmologists are often in error, but never in doubt."

What could await us in a few decades or more? Surely one might anticipate a new theory of cosmology. Much of the observed universe already fits into place in the grand scheme of things. We have come a long way since Newton in terms of our understanding of the physical world. However, there is one element in common with Newton's worldview that ought to give us reason for doubting that we know all of the essential physics. Newton brilliantly foresaw as a consequence of gravity the creation of an almost infinite number of heavenly bodies. The force of gravity inevitably led to instability, and planets and stars had to emerge. But Newton could not understand in terms of known physics why some objects were luminous, like stars, but others, like planets, were not. Newton appealed to a greater authority to solve his problem. The physics breakthrough only emerged 250 years later when Hans Bethe realized what it took to make stars shine, namely thermonuclear energy.

Detection of dark matter poses one of the most urgent problems. It is everywhere, confirmed by several independent types of observations. But we have not detected a single candidate particle. This situation has led some astronomers to question the very basis of gravity. Perhaps the gravity law needs to be modified. Then we can perhaps dispense with dark matter entirely.

Nor is dark energy in any better state. Suppose we dismiss the evidence from supernovae for the acceleration of the universe. After all, we lack a theoretical model of supernovae. No one could disagree that using a measuring tool whose theoretical basis is uncertain is a high-risk procedure. There could be evolutionary systematics that mimic dimming of the most distant supernovae. Maybe intergalactic dust does the dimming. Then we would not need to appeal to acceleration.

We know that something dominates the mass-energy density. Two-thirds of it is not gravitating matter that affects the large-scale distribution of the galaxies. It could be something completely uniform—for example, a new energy field. But this seems an even more radical explanation than Einstein's cosmological constant, which fits all current data.

The situation is curiously reminiscent of our discussions about the multiverse and the anthropic principle. Perhaps we simply lack the correct theory of physics that can cope with the extreme conditions at the beginning of the universe. There are simply too many questions that arise in our present description of the universe. Hiding within these questions could well be the seeds that give rise to a future theory.

Fortunately, we are not in a hurry. Billions of years of exploration lie ahead, provided, of course, that humanity does not self-destruct before then. The sun will live for another 5 billion years. It has barely reached middle age by stellar standards. Much that we cannot even dream of may be achievable on such a timescale. We are on the verge of finding hordes of earthlike planets. Many of these will be older than the earth. With a billion years head start over Homo sapiens, and over our civilization, one cannot begin to imagine what a civilization in some distant galaxy unknown to us might be capable of fabricating. Of course, this does assume that the evolution of life is not improbable given the right environment. We will never be sure of the answer until evidence is found.

Our exploration of the universe need not necessarily be limited in space and time.

We may find a nearby wormhole or naked singularity and exploit it as a time machine. We may develop the technology to be able to travel to other universes. We will see for ourselves how the universe began, and no doubt we will come to understand the nature of dark energy and dark matter, the two great enigmas of our epoch.

Glossary

accretion The process of a rotating central mass attracting gas that orbits around it, as seen in star formation.

angular momentum The law that a spinning body shrinks to keep a constant mass as it accelerates. Tops spin this way.

anthropic principle The idea that the universe produced a brief period when human observers could exist. The "weak" version makes this case only for our cosmic region. The "strong" argues that the universe was destined to produce observers.

antimatter Particles that are negatively identical to corresponding particles and cause mutual annihilation. Fortunately for a universe, more particles than antimatter exist.

baryons Generic name for protons, neutrons, and electrons, the basic constituents of chemical elements. Helium, the lightest element, accounts for a third of all baryonic matter.

big bang A cosmological model postulating that the universe began as a hot explosion and since has been expanding, thinning out, and cooling.

binary galaxies Two galaxies orbiting each other.

blackbody radiation A state of heat that is completely mixed and uniform, like a perfect furnace.

black hole A region where gravity has collapsed and mass is so dense that even light cannot escape.

boundary conditions The earliest conditions of the universe that scientists try to determine to understand its evolution.

chemical enrichment The process of stars producing heavier chemical elements and exploding them into interstellar space.

closed universe A universe with a finite expansion that will collapse back on itself due to gravity and curved space.

cluster The gravitational tendency of galaxies to bunch together rather than spread evenly. Superclusters are larger versions.

compactification How many dimensions of space compact into fewer dimensions. A concept to explain how many dimensions exist in the three known to science.

cosmic background radiation The uniform heat left over from the big bang.

cosmological constant A concept introduced by Einstein to explain a counterforce that stops the universe from collapsing under gravity.

cosmological principle The theoretical assumption that on large scales the universe is homogeneous (looks the same everywhere) and isotropic (looks the same in all directions).

critical density The minimum density required to make a universe that will collapse in the future.

curvature The non-Euclidean geometrical model of the shape of the universe. Angles in curved space are measured differently.

dark energy The repulsive energy thought to be accelerating the expansion rate of the universe. Originally theorized as a cosmological constant by Einstein.

dark matter Unseen matter that is detected only by its gravitational influence. It is believed to hold galaxies together.

elliptical galaxy A galaxy shaped like an amorphous spherical bulge.

Euclidean geometry The measure of "flat" lines and angles.

false vacuum A patch of space that appears "empty" but contains stored energy.

flat universe A universe that is in balance between open (expansion is stronger) and closed (gravity is stronger). The geometry is of an infinite flat plane.

fluctuation Any statistical deviation from uniform conditions, as seen in the early universe's density fluctuations or vacuum fluctuations in empty space.

four fundamental forces The strong and weak nuclear forces, electromagnetism, and gravity.

fundamental constants Basic quantities in physics that give nature its constant qualities.

galactic halo Unseen mass around a galaxy, presumed to be dark matter whose gravity holds it together.

general relativity Einstein's 1916 theory of large-scale gravity curving space.

grand unification The intensely hot instant after the big bang when the four forces of the universe still were combined.

gravity Thought to be a wave or energy field that produces attraction between masses, making possible orbits and the cohesion of a universe.

gravitational lensing Using the visually curved distortion of nearby galaxies to judge the distance of farther-away galaxies.

homogeneity The concept that any large volume of the universe looks the same everywhere.

horizon The edge of the universe as it expands. The distance traversed by a ray of light since the big bang and the limit of human perception of the universe.

Hubble constant A single number that estimates the speed at which a distant galaxy is receding. Calculated by the Hubble law.

Hubble law The statement that a galaxy's recessional speed is proportional to its distance from the observer. It recedes faster as distance increases.

inflationary universe A theory modifying the big bang model by considering an additional instant of inflation that vastly increased the size of the early universe and made its contents uniform.

interstellar medium Gas and dust filling space between stars.

isotropy The assumption that the universe looks the same in all directions from any given point.

large-scale structure The universe's widest distribution of galaxies and other mass forms, defying perfect homogeneity.

metals In astronomy, all elements heavier than hydrogen and helium.

nucleosynthesis The process in stars that turns light nuclei into heavy nuclei, as when hydrogen and helium become oxygen, carbon, and iron.

Olbers' paradox A statement relating to the observation: why is the night sky dark if there are infinite stars and galaxies? The universe's expansion seems to explain why only the light of nearby galaxies reaches earth.

open universe A universe that expands forever since pressure exceeds total gravity in the universe.

parsec The basic measure for galaxy distance, being about 19.2 trillion miles, or 3.26 light years. A kiloparsec is a thousand parsecs, and a megaparsec is a million parsecs.

particle physics The study of the basic particles and forces of nature.

phase transition A marked and sudden change in a physical state, such as when liquid water freezes.

Planck constant A measure of the smallest magnitude of quantum energy. The basis for the Planck density and instant at the time of the big bang.

quantum cosmology Study of the first 10^{-43} seconds after the big bang when quantum mechanics and gravity interacted.

quantum gravity A sought-after theory that explains the unity of quantum relativity in particles and large-scale gravity that curves space.

quantum theory The theory that particles have both wavelike and particle-like qualities, and that a particle's exact speed and location cannot be known at the same moment. The basis of quantum uncertainty.

quasars The universe's brightest objects for their small size, probably intense explosions next to black holes.

redshift How light from receding galaxies moves into the red spectrum when detected by scientific instruments.

scalar field An energy field that moves in all directions because there is no preferred direction.

second law of thermodynamics The statement that isolated systems move toward disorder over time.

singularity A location in time and space thought to be infinitely dense and probably located in the center of a black hole.

special relativity Einstein's 1905 theory that measuring time and length depends on the relative location of the observer.

spectroscopy Study of the wavelengths (by color) of light emitted or reflected by objects.

spectrum The measure of amounts of light in each wavelength, with each having a characteristic color.

spiral galaxy A galaxy shaped like a flat disk with spiral arms and central bulge—for example, the Milky Way.

steady state model A universe that does not change on its largest scales. Expansion is explained by appearance of new particles.

strings The basic multidimensional constituents of matter according to the new physics called string theory.

supernova The brightest kind of exploding star, which is constant in its physics and therefore helpful in judging cosmic distances.

symmetry The quality of staying the same during transformations, as in matter and energy during the first instant after the big bang.

symmetry breaking An irregular separation of components in a symmetry system, usually due to cooling.

vacuum A state of being "empty" or having minimum energy. Quantum uncertainty allows that a vacuum may produce energy fluctuations.

WIMP Weakly interacting massive particles.

wormhole A tunnel to another universe created by a black hole.

Bibliography

Barrow, J. D., P. W. C. Davies, and C. L. Harpereds. *Science and Ultimate Reality: Quantum Theory, Cosmology, and Complexity*. Cambridge: Cambridge University Press, 2004.

Barrow, John, and Frank Tipler. *The Anthropic Cosmological Principle*. Oxford: Oxford University Press, 1986.

Carr, Bernard. *Universe or Multiverse?* Cambridge: Cambridge University Press, 2007.

Carroll, William E. "Big Bang Cosmology, Quantum Tunneling from Nothing, and Creation." *Laval théologique et philosophique* 44, no. 1 (February 1988): 59–75.

Carter, Brandon. "The Significance of Numerical Coincidences in Nature." Unpublished manuscript, 1967. Available at http://arxiv.org/archive/astro-ph, arXiv:0710.3543.

Clayton, Philip. *God and Contemporary Science*. Edinburgh: Edinburgh University Press, 1997.

Craig, William Lane, and Quentin Smith. *Theism, Atheism and Big Bang Cosmology*. Oxford: Clarendon Press, 1993.

Davies, Paul. *The Goldilocks Enigma: Why Is the Universe Just Right for Life?* London: Penguin, 2007.

———. *How to Build a Time Machine*. New York: Viking, 2002.

———. *The Mind of God: Science and the Search for Ultimate Meaning*. London: Penguin, 1993.

Deutsch, D. *The Fabric of Reality*. London: Allen Lane, 1997.

Hawking, Stephen J. *The Universe in a Nutshell*. New York: Bantam Books, 2001.

Heller, Michael. "Cosmological Singularity and the Creation of the Universe." *Zygon* 35, no. 3 (September 2000): 665–685.

———. *Questions to the Universe: Ten Lectures on the Foundations of Physics and Cosmology*. Tucson, AZ: Pachart Publishing House, 1986.

Hoyle, F., G. Burbidge, and J. V. Narlikar. *A Different Approach to Cosmology: From a Static Universe through the Big Bang towards Reality*. Cambridge: Cambridge University Press.

Linde, Andrei. "The Inflationary Multiverse." In *Universe or Multiverse,* edited by Bernard Carr, 127–49. Cambridge: Cambridge University Press, 2007.

Peacock, John A. *Cosmological Physics*. Cambridge (Cambridge: Cambridge University Press, 1999.

Poe, Edgar Allen. *Eureka: A Prose Poem*. 1848.

Rees, Martin. *Just Six Numbers: The Deep Forces That Shape the Universe*. New York: Basic Books, 2000.

———. *Our Cosmic Habitat*. Princeton, NJ: Princeton University Press, 2001.

Ronan, C. A. T*heir Majesties' Astronomers*. London: Bodley Head, 1967.

Russell, Robert J., Nancey Murphy, and C. J. Isham, eds. *Quantum Cosmology and the Laws of Nature: Scientific Perspectives on Divine Action*. Vatican City: Vatican Observatory, 1993.

Russell, Robert J., Nancey Murphy, and Arthur R. Peacocke, eds. *Chaos and Complexity: Scientific Perspectives on Divine Action*. Vatican City: Vatican Observatory and Berkeley, CA: Center for Theology and the Natural Sciences, 2000.

Russell, Robert J., Nancey Murphy, and William R. Stoeger, SJ, eds., *Scientific Perspectives on Divine Action: Twenty Years of Challenge and Progress*. Vatican City State: Vatican Observatory Publications, Berkeley: The Center for Theology and the Natural Sciences; Notre Dame, IN: University of Notre Dame Press, 2007.

Saunders, Nicholas. *Divine Action and Modern Science*. Cambridge: Cambridge University Press, 2002.

Seneca, Lucius Annaeus. *Natural Questions,* book 7. Quoted in Carl Sagan, *Cosmos*. New York: Random House, 1980.

Silk, Joseph. *The Big Bang*. 3rd ed. London: W. H. Freeman & Co., 2001.

———. *The Infinite Cosmos: Questions from the Frontiers of Cosmology*. New York: Oxford University Press, 2006.

———. *On the Shores of the Unknown: A Short History of the Universe*. Cambridge: Cambridge University Press, 2004.

Smolin, Lee. *The Life of the Cosmos*. Oxford: Oxford University Press, 1997.

———. *Three Roads to Quantum Gravity*. London: Weidenfeld & Nicolson, 2000.

———. *The Trouble with Physics*. London: Allen Lane, 2007.

Steinhardt, Paul, and Neil Turok. *The Endless Universe: Beyond the Big Bang.* London: Weidenfeld & Nicolson, 2007.

Susskind, Leonard. *The Cosmic Landscape: String Theory and the Illusion of Intelligent Design.* New York: Little, Brown, 2005.

Vilenkin, A. *Many Worlds in One: The Search for Other Universes.* New York: Hill & Wang, 2006.

Wheeler, J. A. *A Journey into Gravity and Spacetime.* New York: Scientific American Library, 1990.

Zee, A. *Quantum Field Theory in a Nutshell.* Princeton, NJ: Princeton University Press, 2003.

Index

acceleration, 29, 119, 124
accretion, 55–56
age: earth's, 128; Hubble's constant and, 51; of stars, 12, 57; of universe, 17–18, 51, 149–51, 155; velocity dispersion/iron abundance and, 61
Alpher, Ralph, 21
anthropic principle, 160; cosmic evolution and, 18; Grand Designer hypothesis and, 163, 168; observer regarding, 170–71; physics and, 161, 163; random chance and, 163–64; strong version of, 162; weak version of, 161–62
antigravity constant, 7, 118, 122
Arrhenius, Svante, 147–48
atmospheric smearing, 140

background fluctuations: acoustics concerning, 122–23; big bang theory and, 26–29; bottom-up evolution and, 42–43; COBE satellite verifies, 49, 49n6; concept of, 26–27; dark matter concerning, 43–44, 48–49; frozen, 42; galaxy formation and, 27–29, 46–47; gravitation regarding, 26; horizon regarding, 40–41; measurement of, 40, 46–50; microwave experimentation of, 48; predictions of, 47, 49; quantum theory concerning, 38; theories for testing, 47–48
background radiation: big bang verification by, 22–25; blackbody radiation and, 23–25; COBE satellite and, 24–25; discovery of, 22–23. *See also* background fluctuations; cosmic microwave background
bar, 10–11
barn, 108
baryons: black holes and, 114–15; cosmic web and, 76; dark matter v., 45–46, 66–67; density regarding, 155; elements and, 27, 27n2; elliptical galaxies and, 97–98; galaxy formation and, 43–46, 93–94; gravity and, 72–73; location of, 66–67; matter budget regarding, 66; Milky Way regarding, 93–94; oscillations of, 122–23
Bayes' theorem, 39
Bayes, Thomas, 39, 173
Bethe, Hans, 55
big bang theory: acceleration and, 29; blackbody and, 23–25; COBE satellite and, 24–25; cold v. hot, 20, 22, 23; evidence for, 18–19; expansion verifying, 19;

flat universe and, 30; fluctuations verifying, 26–29; Gamow regarding, 19–23; light elements verifying, 19–22; naming of, 23; particle physics' influence on, 30–31; Penzias/Wilson and, 22–23; radiation verifying, 22–25; relics of, 27; steady state hypothesis v., 144–46; thermodynamics and, 24. *See also* expanding universe theory; inflationary cosmology
black holes: binary, 117; cosmic evolution and, 13; cosmic origin and, 17; formation of, 114–15; intermediate-mass, 116–17; mass range of, 115; merging of, 115–16; short-lived mini, 166; spikes and, 114–17; spinning, 181–82; star formation producing, 60; time travel via, 181–83; universe creation via, 164–66. *See also* supermassive black holes
blackbody radiation, 23–25
Bohr, Niels, 135
Bond, Dick, 49n4
Bond, Hermann, 145
Born, Max, 135
bottom-up evolution, 42–43
Brownian motion, 133
Bruno, Giordano, 132
Burbidge, Margaret, 147

Calabi-Yau manifolds, 176–77
caustics, 74–75
Chandrasekhar, Subhramanian, 60
clumps, dark matter, 108–9, 113–14
clusters, galaxy, 63–64, 125–26
clusters, star, 57–58, 62
COBE. *See* Cosmic Background Explorer satellite
coincidence, 18
cold dark matter paradigm, 74
cold theory, 20
COMPTON (gamma ray observatory), 112–13
constant, cosmological. *See* cosmological constant
Cosmic Background Explorer (COBE) satellite, 24–25, 49, 49n6
cosmic evolution: black holes/quasars and, 13; coincidence regarding, 18; galaxies and, 13, 14, 42–43, 45–46; Planck's density and, 16–17; recycling process in, 14–15; science/religion and, 17–18; solar system and, 58; star death and, 15; star formation in, 13–14; stars and, 13–14, 15, 58–61, 89–90. *See also* big bang theory; expanding universe theory; inflationary cosmology
cosmic microwave background: acoustic fluctuations in, 123; Hoyle concerning, 146; primeval atom theory regarding, 9; WMAP and, 178–79. *See also* background fluctuations; background radiation
cosmic web: baryons and, 76; caustics regarding, 74–75; cold dark matter paradigm and, 74; galaxy surveys/simulations and, 75
cosmological constant, 118; changes in, 129–31; as dark energy, 122; Extremely Large Telescope and, 131; Hubble's, 8–9, 51; issues surrounding, 127–28; late development/smallness of, 128; Lemaître and, 7; multiple universes and, 129; Oklo Uranium mine and, 130
cosmology: cosmologist error in, 184–85; history regarding, 3, 134–35; pathways of, 149; seven numbers regarding, 155–56

critical density, 67
curved space, 50–51

dark energy: acceleration and, 29, 119, 124; baryon oscillations measuring, 122–23; Bayes' method and, 173; cluster counts revealing, 125–26; density regarding, 122, 155–56; eternal inflation and, 166–67; experiments revealing, 124–26; flat universe and, 30, 118–19; galaxy formation and, 174–75; gravity concerning, 29, 124–25; hunt for, 122–27; inflation field and, 173–74; inflationary cosmology and, 118–19, 185–86; matter budget regarding, 66; observational bias concerning, 172; pressure/density ratio regarding, 122; puzzles, 121–22; redshift surveys and, 123–24; string theory regarding, 172–73; supernova hunting and, 124; telescopes and, 126–27; theory of, 118–19, 185–86; time travel via, 182. *See also* cosmological constant
dark matter, 154–55; antimatter regarding, 103; background fluctuations and, 43–44, 48–49; balloon experiment on, 111–12; baryons and, 45–46, 66–67; black holes/spikes and, 114–17; clumps, 108–9, 113–14; clusters indicating, 63–64; cold paradigm of, 74; COMPTON/FERMI and, 112–13; density regarding, 155–56; direct detection of, 106–10; flat space/universe and, 30, 69; galaxies and, 63–66; gamma rays and, 111–13; gravity and, 64; indirect detection of, 110–13; in inflationary cosmology, 43–46; Large Hadron Collider and, 105–6; large scale measurement of, 68–70; Lee/Weinberg concerning, 103–4; lensing indicating, 65–66; LIBRA and, 109; light bending, 124–25; location of, 102–3; matter budget regarding, 66; in Milky Way, 102; neutralino and, 105; PAMELA and, 112; positrons concerning, 110, 111–12; rotation curves indicating, 64–65; SMBHs and, 115, 116; solutions, 72–73; sun regarding, 109–10; supersymmetry and, 104–5; underground detection of, 107–8, 109; weak interaction/particles of, 46, 102, 103–7; WIMPs as, 106; X-ray studies of, 125
Dicke, Robert, 24, 162
dimensions, multiple, 130
Dirac, Paul, 142
disk galaxies, 10; cold gas and, 85–86; computer modeling of, 87–88; dark matter and, 64–65; feedback regarding, 91–92, 93; gas ejection in, 94–95; gas infall and, 85; inefficiency in, 83–84, 91–96; low-surface-brightness dwarfs and, 95–96; Milky Way and, 93–95; rotation of, 84–85, 87; rotational velocity/stellar mass and, 86–87; self-regulation in, 92–93; star age regarding, 12; star formation in, 83–88; supernova heating in, 94
Doppler effect, 6
dry mergers, 88–89
dwarf galaxies problem, 76–78

earth, 128
Eddington, Arthur: background of, 138; cosmological constant and, 118; eclipse experiments and, 139–40; Einstein on, 143;

general relativity and, 138–40; Hutchinson on, 142–43; inflation and, 30; Ronan on, 142; on star formation, 54; theory of everything by, 141–42; theory/observation maxim of, 141
Einstein, Albert: antigravity constant of, 7, 118, 122; background of, 133–34; determinism v. probability and, 136; on Eddington, 143; Friedmann and, 4; general relativity and, 134; gravity theory and, 3, 7, 118, 122, 134; on imagination, 132; mistake of, 136–37; on morality, 137–38; on nuclear energy, 184; quantum mechanics and, 135–36, 184; on relativity, 135; religion and, 137; on simplicity, 133; unified field quest of, 135–37
elements, 20–22
elliptical galaxies, 10–11; baryonic shortfall concerning, 97–98; chemistry concerning, 89–90; dark matter and, 65; formation theory of, 88–90, 96–97; mergers and, 88–89; observation regarding, 90, 97–98; SMBHs and, 90–91; star age regarding, 12; star formation in, 88–91; temperature regarding, 96–97
entropy. *See* thermodynamics, second law of
eternal inflation, 166–67
Euclidian geometry, 39
Euler number, 176–77
evolution. *See* cosmic evolution
expanding universe theory: acceleration concerning, 29; early work on, 3–9; evidence for, 18–19; expansion rates in, 8–9, 17; Friedmann and, 3–4; Hubble concerning, 5–7, 16; Hubble diagram and, 17; Hubble's law illuminating, 6–7; Lemaître regarding, 4–5; Planck's density and, 16–17; *The Realm of the Nebulae* on, 7; Slipher and, 6; time dilation test and, 8. *See also* big bang theory; dark energy; inflationary cosmology
Extremely Large Telescope, 131

false vacuum, 128
feedback: challenges regarding, 100–101; disk galaxies and, 91–92, 93; mergers regarding, 98–99; SMBHs and, 99–100; supernova creating, 100
FERMI. *See* Gamma Ray Large Area Space Telescope
fingers of God, 68
first acoustic peak, 49n5
Fisher, Richard, 86
flat space: big bang and, 30; confirmation of, 50–51; dark energy producing, 30, 118–19; dark matter and, 30, 69; Hubble's constant and, 51; probability of, 39
fluctuations. *See* background fluctuations
Fowler, William, 144, 147
Friedmann, Alexander, 3–4

galaxies: background fluctuations and, 27–29, 46–47; background on, 10; bottom-up evolution and, 42–43; clusters of, 63–64, 125–26; colors of, 29; dark matter and, 63–66; discovery of, 5; dwarf problem of, 76–78; evolution of, 13, 14, 42–43, 45–46; first, 45; formation of, 11, 27–29, 42–47, 80, 83–84, 93–94, 174–75; halos and, 44; Hubble's types of, 11–12; initial conditions regarding, 46–47; life

spans of, 14; prediction regarding, 71–72; radio, 145; redshift measurement and, 4, 6–7; shapes of, 10–11, 29; size/mass limits of, 78; star age and, 12; star formation in, 82–83; surveys/simulations of, 75. *See also* disk galaxies; elliptical galaxies; massive galaxy problem; spiral galaxies
Gamma Ray Large Area Space Telescope (FERMI), 112–13
gamma rays: clumps and, 114; dark matter and, 111–13
Gamow, George: background on, 19; big bang theory and, 19–23; hot origins and, 20, 22; Hoyle and, 146–47; light elements and, 19–22, 146–47, 152
general relativity: atmospheric smearing and, 140; eclipse verification of, 139–40; Eddington and, 138–40; Einstein's, 134
globular star clusters. *See* clusters, star
Gold, Tommy, 145
Goldilocks enigma, 168
googol, 34–35, 35n3
Grand Designer hypothesis, 163, 168
grand unification, 32–34, 154
gravity: antigravity and, 7, 118, 122; background fluctuations and, 26; baryons influenced by, 72–73; dark energy and, 29, 124–25; dark matter and, 64; dwarf galaxies problem and, 77; Einstein concerning, 3, 7, 118, 122; general relativity and, 134; human size regarding, 161–62; lensing, 65–66; light bending via, 124–25; LISA and, 36; quantum theory and, 35, 136; star formation and, 52–53; symmetry breaking and, 32–33. *See also* general relativity
Guth, Alan, 35

Hall, Lawrence, 169
Halley, Edmond, 151
halos, 44
Harrison, Edward, 178
Hartle, James, 170
helioseismology, 109
Herman, Robert, 21
history: chemical, 62; cosmology founders in, 3, 134–35; redshift, 4–5
horizon, 40–41, 42
Hoyle, Fred: background on, 143; on big bang theory, 23; chemical elements origins and, 143–44, 146–47; cosmic microwave background and, 146; Gamow and, 146–47; panspermia theory regarding, 147–48; steady state hypothesis of, 144–46; as unconventional, 144
Hubble diagram, 17, 119
Hubble, Edwin: background on, 5; constant of, 8–9; expansion regarding, 5–7, 16; galaxy classification by, 11–12; Humason and, 5–6; redshift measurement and, 4, 6–7; de Sitter and, 16; tired light theory of, 8
Hubble Space Telescope, 127
Hubble's constant: accepted value of, 9; debates surrounding, 8–9; flat/curved space and, 51
Hubble's law, 5, 6–7, 121
human beings, 160–62
Humason, Milton, 5–6
Hutchinson, I.H., 142–43
Huxley, Thomas H., 158
hydrogen, 21
hypernovae, 60–61

inflationary cosmology: bottom-up evolution in, 42–43; dark energy theory and, 118–19, 185–86; dark matter in, 43–46; Eddington and, 30; eternal inflation and, 166–67; flat space and, 39; fluctuation measurement in, 40, 46–50; forces involved in, 32; grand unification and, 32–34, 154; horizon regarding, 40–41, 42; inflation field and, 173–74; Large Hadron Collider and, 35–36; limitations of, 35–37; LISA and, 36; origins of, 30–31; particle physics and, 30–37; quantum theory and, 38; signatures of, 36–37; successes of, 178–79; symmetry breaking and, 31–35; theory variety in, 38–39; time measurement in, 34–35; universe's uniformity and, 33, 34. *See also* big bang theory; expanding universe theory

Japanese Subaru telescope, 126
Jeans, James, 53–54
Jones, Harold Spencer, 184–85

Kant, Immanuel, 5
Kennicutt, Robert, 85
Kepler, Johannes, 150

Large Hadron Collider, 152; dark matter and, 105–6; inflationary cosmology and, 35–36; supersymmetry regarding, 104
Lee, Ben, 103–4
Lemaître, Abbé, 17
Lemaître, Georges, 3; background on, 4; cosmological constant and, 118; expanding universe test of, 4–5; primeval atom theory of, 7, 9
lensing, 65–66
LIBRA (dark matter experiment), 109
Linde, Andrei, 35
LISA (gravity wave detector), 36
low-surface-brightness dwarfs, 95–96
Lundmark, Knut, 6

Magorrian, John, 90
massive galaxy problem: black hole feeding and, 79–80; gas clouds concerning, 81–82; quasar outflows and, 80–81; size/mass limits and, 78; SMBHs in, 79–82; star formation and, 78–79
mathematics, 175–77
matter budget, 66
mergers: feedback and, 98–99; types of, 88–89
microwave background. *See* cosmic microwave background
Milky Way: acceleration regarding, 119; black holes in, 115–16; chemistry of, 61–62; dark matter in, 102; as disk example, 93–95; gamma rays in, 112; shape of, 10–11; star formation in, 29, 84; velocity dispersion and, 61
minor merger, 89
Mount Wilson Observatory, 4, 5
multiple dimensions. *See* dimensions, multiple
multiverse: frontier limits of, 169; Goldilocks enigma and, 168; theories of, 167–68; time travel and, 179–83

Narlıkar, Jayant, 145
Natural Questions (Seneca), 183
natural selection theory, 164–66
neutralino, 105
neutron star, 59–60
Newton, Isaac: Einstein on, 134; on star/planet formation, 52–53

Oklo Uranium mine, 130
Olbers, Heinrich Wilhelm, 150

Page, Don, 170
PAMELA (space satellite), 112
panspermia theory, 147–48
parsec, 9n1
particle accelerator, 106. *See also* Large Hadron Collider
particle physics: antimatter and, 103; big bang theory and, 30–31; dark energy puzzle and, 121–22; dark matter, 46, 102–17; grand unification and, 32–34; inflationary cosmology and, 30–37; Large Hadron Collider and, 35–36, 104, 105–6, 152; positrons and, 110, 111–12, 114; proton decay and, 153–54; quarks/gluons in, 32; string theory and, 175; supersymmetry and, 104; symmetry breaking and, 31–35; weak interaction in, 46, 102, 103–7; WIMPs in, 106, 109, 113–14
Penzias, Arno, 22–23, 25, 152
photoelectric effect, 133, 135
Pius XII (pope), 17
Planck density, 16–17
Planck instant, 33, 35, 35n3
Planck, Max, 16
planet formation, 52–55
planetary nebula, 15
Plummer, H. C., 141
Poe, Edgar Allan, 151
positrons: clumps and, 114; dark matter concerning, 110, 111–12
primeval atom theory, 7, 9
proton decay, 153–54

quantum gravity, 35, 136
quantum mechanics: Einstein and, 135–37, 184; photoelectric effect, 133, 135
quantum theory, 38
quark-gluon plasma, 32
quasars: cosmic evolution and, 13; massive galaxy problem and, 80–81

radio galaxies, 145
The Realm of the Nebulae (Hubble), 7
recession velocity, 19
red giant, 58–59
red supergiant, 59
redshift: cosmological history and, 4–5; dark energy surveys and, 123–24; measurement, 4, 6–7; de Sitter and, 16; Slipher and, 6
relativity, 134–35. *See also* general relativity
religion: Einstein and, 137; universe age and, 17–18
Robertson, J. P., 4
Ronan, C. A., 142
rotation curves, 64–65
Rumsfeld, Donald, 122
Ryle, Martin, 145

Sachs, Rainer, 47n2
Salpeter, Edwin, 146
Sandage, Alan, 19
Schmidt, Martin, 85
Schmidt-Kennicutt law, 85–86
Schrödinger, Erwin, 171
Schrödinger's cat, 171
self-regulation, 92–93
Seneca, Lucius Annaeus, 183
Silk, Joe, 49n4
Sitter, Wilhem de, 16
Slipher, Vesto, 6
SMBH. *See* supermassive black holes
Smolin, Lee, 164
solar system, 58–59
sound speed, 55–57
special relativity, 133–34
spectroscopy, 6
speed of sound. *See* sound speed
spheroidal galaxies. *See* elliptical galaxies

spikes, dark matter, 114–17
spiral galaxies, 10–11, 83–84. *See also* Milky Way
Square Kilometer Array, 126–27
Srednicki, Mark, 170
standard candles, 19
stars: accretion by, 55–56; ages of, 12, 57; black holes and, 60; chemistry evolution via, 58–61, 89–90; clusters of, 57–58, 62; cool universe regarding, 56–57; death of, 15; discs and, 83–88; early theories on, 52–54; Eddington on, 54; elliptical galaxies and, 88–91; explosion/enrichment process in, 56; first, 45, 55–61; formation of, 13–14, 29, 45, 52–62, 78–79, 82–91; galaxy formation and, 11; Hoyle concerning, 143–44; hypernovae and, 60–61; inefficiency regarding, 83–84; Jeans on, 53–54; massive galaxy problem and, 78–79; Milky Way and, 29, 84; neutron/white dwarf, 59–60; Newton on, 52–53; planet formation v., 52–55; red giant, 58–59; red supergiant, 59; solar system creation and, 58; sound speed influencing, 55–57; supernovas and, 59–61; velocity dispersion, 61. *See also* sun
steady state hypothesis: density issue regarding, 145–46; Hoyle's, 144–46; microwave background and, 146; radio galaxies and, 145
string theory, 136; dark energy and, 172–73; mathematics regarding, 175–77
sun: dark matter and, 109–10; fuel of, 55; future of, 58–59; temperature and, 109
supermassive black holes (SMBH): dark matter regarding, 115, 116; elliptical galaxies and, 90–91; feedback and, 99–100; feeding of, 79–80; galaxy formation and, 80; gas clouds and, 81–82; massive galaxy problem and, 79–82; quasar outflows regarding, 80–81
supernova: acceleration regarding, 119, 121; distance measurement using, 119–21; dwarf galaxies problem and, 76–77; feedback and, 100; heating, 94; hunting, 124; Ia, 120; precision breakthrough regarding, 120–21; star formation and, 59–61
symmetry: breaking, 32–33, 154; inflationary cosmology and, 31–35; super, 104–5

telescope: dark energy and, 126–27; Extremely Large, 131; FERMI, 112–13; Hubble, 127; Japanese Subaru, 126
Their Majesties' Astronomers (Ronan), 142
theory, limitations of, 71–72
thermodynamics, second law of, 24
Third Texas Symposium on Relativistic Astrophysics, 23
time dilation, 8
time travel: possibility of, 179–80; recipe for, 181–83; spaceship for, 182; space/time tunnel and, 180–82; spinning black hole for, 181–82; wormhole technology for, 180–83
tired light, 8
true vacuum, 128
Tully, Brent, 86
Tully-Fisher law, 86–87

unified field, 135–37
universe(s): age of, 17–18, 51, 149–51,

155; beginning of, 152–53, 175–77; cool, 56–57; density of, 152; distance/time measurement of, 149–51; dynamism of, 151; eternal inflation and, 166–67; finite v. infinite, 157–59; flat v. curved, 30, 118–19; frontier limits and, 169; future of, 156–57; hot, 55–56; infinite, 168; mathematics and, 175–77; multiple, 129, 167–68; natural selection theory of, 164–66; observer regarding, 170–71; as particle accelerator, 106; seven numbers and, 155–56; static, 16; uniformity of, 33, 34. *See also* big bang theory; expanding universe theory; inflationary cosmology
Uson, Juan, 48n3
Ussher, James, 17

vacuum. *See* false vacuum; true vacuum
velocity dispersion, 61
Vilenkin, Alex, 162
Vittorio, Nicola, 49n4

weakly interacting massive particles (WIMPs): clumps regarding, 113–14; dark matter as, 106; detection of, 109
Weinberg, Steven, 103–4, 173
wet mergers, 88
white dwarf, 59–60, 120
White, T.H., 158
wiggles, 69
Wilkinson, David, 48n3
Wilkinson Microwave Anisotropy Probe (WMAP) satellite, 178–79
Wilson, Robert, 22–23, 25, 152
WIMPs. *See* weakly interacting massive particles
WMAP. *See* Wilkinson Microwave Anisotropy Probe satellite
Wolfe, Art, 47n2
Wooley, Richard, 184–85
wormhole: black hole and, 181–83; space ship and, 182; space/time tunnel and, 180–82; technology, 180–83

X-rays, 125

Zel'dovich, Yaakov, 178; caustics and, 74–75; cold theory and, 20; on origins model, 7